2019 SQA Specimen and Past Papers with Answers

National 5

GEOGRAPHY

SQA
National 5 GEOGRAPHY

2018 & 2019 Exams
and 2017 Specimen Question Paper

Hodder Gibson Study Skills Advice – National 5 Geography – page 3
Hodder Gibson Study Skills Advice – General – page 5
2017 SPECIMEN QUESTION PAPER – page 7
2018 EXAM – page 35
2019 EXAM – page 69
ANSWERS – page 107

This book contains the 2017 Specimen Question Paper for National 5 Geography, and the official SQA 2018 and 2019 Exams, with associated SQA-approved answers modified from the official marking instructions that accompany the paper.

In addition the book contains study skills advice. This advice has been specially commissioned by Hodder Gibson, and has been written by experienced senior teachers and examiners in line with the new National 5 syllabus and assessment outlines. This is not SQA material but has been devised to provide further guidance for National 5 examinations.

Hodder Gibson is grateful to the copyright holders for permission to use their material. Every effort has been made to trace the copyright holders and to obtain their permission for the use of copyright material. Hodder Gibson will be happy to receive information allowing us to rectify any error or omission in future editions.

Hachette UK's policy is to use papers that are natural, renewable and recyclable products and made from wood grown in well-managed forests and other controlled sources. The logging and manufacturing processes are expected to conform to the environmental regulations of the country of origin.

Orders: please contact Bookpoint Ltd, 130 Park Drive, Milton Park, Abingdon, Oxon OX14 4SE. Telephone: (44) 01235 827827. Fax: (44) 01235 400454. Lines are open 9.00–5.00, Monday to Friday, with a 24-hour message answering service. Visit our website at www.hoddereducation.co.uk.
If you have queries or questions that aren't about an order, you can contact us at hoddergibson@hodder.co.uk

This collection first published in 2019 by
Hodder Gibson, an imprint of Hodder Education,
An Hachette UK Company
211 St Vincent Street
Glasgow G2 5QY

Typeset by Aptara, Inc.

Printed in the UK

A catalogue record for this title is available from the British Library

ISBN: 978-1-5104-7818-3

2 1

2020 2019

Introduction

National 5 Geography

This book of SQA past papers contains the question papers used in the 2018 and 2019 exams (with the answers at the back of the book). A specimen question paper reflecting the content and duration of the revised exam in 2018 is also included. All of the question papers included in the book provide excellent representative practice for the final exams.

Using these papers as part of your revision will help you to develop the vital skills and techniques needed for the exam, and will help you to identify any knowledge gaps you may have.

It is always a very good idea to refer to SQA's website for the most up-to-date course specification documents. These are available at www.sqa.org.uk/sqa/47446

The exam

The course assessment will consist of two parts: a question paper (80 marks) and an assignment (20 marks). The question paper is therefore worth four-fifths of the overall marks of the course assessment, and the assignment one-fifth. The assignment is completed throughout the year and submitted to SQA to be marked in April. Your assignment marks are then added to the marks achieved in the exam paper to give you a final award.

The question paper

The purpose of the question paper is to allow you to demonstrate the skills you have acquired and to reveal the knowledge and understanding you have gained from the topics studied throughout the course. The question paper will give you the chance to show your ability in describing, explaining, matching and evaluating a broad range of geographical information as well as using a variety of maps and demonstrating proficiency in Ordnance Survey (OS) skills. Candidates will complete this question paper in 2 hours and 20 minutes. Questions will be asked on a local, regional and global scale. The question paper has three sections.

Section 1: Physical Environments

This section is worth 30 marks. Candidates will answer a mixture of limited/extended-response questions by using the knowledge, understanding and skills learned throughout the course. In this section there is a choice. Candidates should answer **either** Question 1: glaciation/coasts or Question 2: rivers/limestone. This will be dependent on the subjects taught at your school. Some topics you could be asked to answer questions on include **Weather, Landscape formations** within Scotland and/or the UK, and **Land use management** – conflicts and solutions. In this section you may also be examined on your Ordnance Survey skills using a map.

Section 2: Human Environments

This section is worth 30 marks. As in Section 1, candidates will answer a mixture of limited/extended-response questions by using the knowledge, understanding and skills learned throughout the course. Candidates should answer **all** questions in this section, which will be drawn from both the developed and developing world. Some topics you could be asked questions on include **Population** (development indicators, population distribution, factors affecting birth rates and death rates), **Urban** (land use characteristics in cities in the developed world, recent developments in developed world cities, strategies to improve shanty towns) and **Rural** (changes in rural landscapes in both the developed and developing world). In this section you may also be examined on your Ordnance Survey skills using a map.

Section 3: Global Issues

This section is worth 20 marks, made up of two 10-mark questions. Again, candidates will answer these questions by using the knowledge, understanding and skills learned throughout the course. In this section there is a choice of questions. Candidates should answer **two** questions from a choice of six. Your choice will be dependent on the topics taught at your school. The choice of topics is: **Climate change**, **Natural regions**, **Environmental hazards**, **Trade and globalisation**, **Tourism** and **Health**.

Types of questions

The main types of questions used in the paper are: **Describe, Explain, Give reasons, Match, Give advantages and/or disadvantages,** and **Give map evidence.**

Describe questions

You must make a number of relevant, factual points. These should be key points taken from a given source, for example a map, diagram or table.

Explain or Give reasons questions

You should make a number of points giving clear reasons for a given situation. The command word "explain" will be used when you are asked to demonstrate knowledge and understanding. Sometimes the command words "give reasons" may be used as an alternative to "explain".

Match questions

You are asked to match two sets of variables, for example to match features to a correct grid reference.

Advantages and/or disadvantages questions

You should select relevant advantages or disadvantages of a proposed development, for example the location of a new shopping centre, and demonstrate your understanding of the significance of the proposal.

Give map evidence questions

You should look for evidence on the map and make clear statements to support your answer.

Some tips for revising

- To be best prepared for the examination, organise your notes into sections. Try to work out a schedule for studying with a programme which includes the sections of the syllabus you intend to study.
- Organise your notes into checklists and revision cards.
- Make sure you have a copy of the examination timetable and have planned a schedule for studying.
- Try to avoid leaving your studying to a day or two before the exam. Also try to avoid cramming your studies into the night before the examination, and especially avoid staying up late to study.
- One useful technique when revising is to use summary note cards on individual topics.
- Make use of past paper questions to test your knowledge and skills. Go over your answers and give yourself a mark for every correct point you make when comparing your answer with your notes.
- If you work with a classmate, try to mark each other's practice answers.
- Practise your diagram-drawing skills and your writing skills. Ensure that your answers are clearly worded. Try to develop the points that you make in your answers.

Some tips for the exam

- Do not write lists, even if you are running out of time. You will lose marks. If the question asks for an opinion based on a choice, for example on the suitability of a particular site or area for a development, do not be afraid to refer to negative points such as why the alternatives are not as good. You will get credit for this.
- Arrive at the examination in plenty of time with the appropriate equipment – pen, pencil, rubber and ruler.
- Carefully read the instructions on the paper and at the beginning of each part of the question.
- Answer all of the compulsory questions in each paper you sit.
- Use the number of marks as a guide to the length of your answer.
- Try to include examples in your answer wherever possible. If asked for diagrams, draw clear, labelled diagrams.
- Read the question instructions very carefully. If the question asks you to "describe", make sure that this is what you do.
- If you are asked to "explain", you must use phrases such as "due to", "this happens because" and "this is a result of". If you describe, rather than explain, you will lose most of the marks for that question.
- If you finish early, do not leave the exam. Use the remaining time to check your answers and go over any questions which you have partially answered, especially Ordnance Survey map questions.
- Practise drawing diagrams which may be included in your answers, for example corries or pyramidal peaks.
- Make sure that you have read the instructions on the question carefully and that you have avoided needless errors. For example, answering the wrong sections or failing to explain when asked to, or perhaps forgetting to refer to a named area or case study.
- One technique which you might find helpful when answering 5- or 6-mark questions is to "brainstorm" possible points for your answer. You can write these down in a list at the start of your answer. As you go through your answer, you can double-check with your list to ensure that you have put as much into your answer as you can. This stops you from coming out of the exam and being annoyed that you forgot to mention an important point.

Common errors

Markers of the external examination often remark on errors which occur frequently in candidates' answers. These include the following:

Lack of sufficient detail

- Many candidates fail to provide sufficient detail in answers, often by omitting reference to specific examples, or not elaborating or developing points made in their answer. As noted above, a good guide to the amount of detail required is the number of marks given for the question. If, for example, the total marks offered is 6, then you should make at least six valid points.

Listing

- If you write a simple list of points rather than fuller statements in your answer, you will automatically lose marks. For example, in a 4-mark question, you will obtain only 1 mark for a list.
- The same rule applies to a simple list of bullet points. However, if you couple bullet points with some detailed explanation, you could achieve full marks.

Irrelevant answers

- You must read the question instructions carefully so as to avoid giving answers which are irrelevant to the question. For example, if you are asked to "explain" and you simply "describe", you will lose marks. If you are asked for a named example and you do not provide one, you will forfeit marks.

Repetition

- You should be careful not to repeat points already made in your answer. These will not gain any further marks. You may feel that you have written a long answer, but it may contain the same basic information repeated again and again. Unfortunately, these repeated statements will be ignored by the marker.

Good luck!

Remember that the rewards for passing National 5 Geography are well worth it! Your pass will help you to get the future you want for yourself. In the exam, be confident in your own ability. If you're not sure how to answer a question, trust your instincts and just give it a go anyway – keep calm and don't panic! GOOD LUCK!

Study Skills – what you need to know to pass exams!

General exam revision: 20 top tips

When preparing for exams, it is easy to feel unsure of where to start or how to revise. This guide to general exam revision provides a good starting place, and, as these are very general tips, they can be applied to all your exams.

1. Start revising in good time.

Don't leave revision until the last minute – this will make you panic and it will be difficult to learn. Make a revision timetable that counts down the weeks to go.

2. Work to a study plan.

Set up sessions of work spread through the weeks ahead. Make sure each session has a focus and a clear purpose. What will you study, when and why? Be realistic about what you can achieve in each session, and don't be afraid to adjust your plans as needed.

3. Make sure you know exactly when your exams are.

Get your exam dates from the SQA website and use the timetable builder tool to create your own exam schedule. You will also get a personalised timetable from your school, but this might not be until close to the exam period.

4. Make sure that you know the topics that make up each course.

Studying is easier if material is in manageable chunks – why not use the SQA topic headings or create your own from your class notes? Ask your teacher for help on this if you are not sure.

5. Break the chunks up into even smaller bits.

The small chunks should be easier to cope with. Remember that they fit together to make larger ideas. Even the process of chunking down will help!

6. Ask yourself these key questions for each course:

- Are all topics compulsory or are there choices?
- Which topics seem to come up time and time again?
- Which topics are your strongest and which are your weakest?

Use your answers to these questions to work out how much time you will need to spend revising each topic.

7. Make sure you know what to expect in the exam.

The subject-specific introduction to this book will help with this. Make sure you can answer these questions:

- How is the paper structured?
- How much time is there for each part of the exam?
- What types of question are involved? These will vary depending on the subject so read the subject-specific section carefully.

8. Past papers are a vital *revision tool!*

Use past papers to support your revision wherever possible. This book contains the answers and mark schemes too – refer to these carefully when checking your work. Using the mark scheme is useful; even if you don't manage to get all the marks available first time when you first practise, it helps you identify how to extend and develop your answers to get more marks next time – and of course, in the real exam.

9. Use study methods that work well for you.

People study and learn in different ways. Reading and looking at diagrams suits some students. Others prefer to listen and hear material – what about reading out loud or getting a friend or family member to do this for you? You could also record and play back material.

10. There are three tried and tested ways to make material stick in your long-term memory:

- Practising – e.g. rehearsal, repeating
- Organising – e.g. making drawings, lists, diagrams, tables, memory aids
- Elaborating – e.g. incorporating the material into a story or an imagined journey

11. Learn actively.

Most people prefer to learn actively – for example, making notes, highlighting, redrawing and redrafting, making up memory aids, or writing past paper answers. A good way to stay engaged and inspired is to mix and match these methods – find the combination that best suits you. This is likely to vary depending on the topic or subject.

12. Be an expert.

Be sure to have a few areas in which you feel you are an expert. This often works because at least some of them will come up, which can boost confidence.

13. Try some visual methods.

Use symbols, diagrams, charts, flashcards, post-it notes etc. Don't forget – the brain takes in chunked images more easily than loads of text.

14. Remember – practice makes perfect.

Work on difficult areas again and again. Look and read – then test yourself. You cannot do this too much.

15. Try past papers against the clock.

Practise writing answers in a set time. This is a good habit from the start but is especially important when you get closer to exam time.

16. Collaborate with friends.

Test each other and talk about the material – this can really help. Two brains are better than one! It is amazing how talking about a problem can help you solve it.

17. Know your weaknesses.

Ask your teacher for help to identify what you don't know. Try to do this as early as possible. If you are having trouble, it is probably with a difficult topic, so your teacher will already be aware of this – most students will find it tough.

18. Have your materials organised and ready.

Know what is needed for each exam:

- Do you need a calculator or a ruler?
- Should you have pencils as well as pens?
- Will you need water or paper tissues?

19. Make full use of school resources.

Find out what support is on offer:

- Are there study classes available?
- When is the library open?
- When is the best time to ask for extra help?
- Can you borrow textbooks, study guides, past papers, etc.?
- Is school open for Easter revision?

20. Keep fit and healthy!

Try to stick to a routine as much as possible, including with sleep. If you are tired, sluggish or dehydrated, it is difficult to see how concentration is even possible. Combine study with relaxation, drink plenty of water, eat sensibly, and get fresh air and exercise – all these things will help more than you could imagine. Good luck!

NATIONAL 5

2017 Specimen Question Paper

National
Qualifications
SPECIMEN ONLY

S833/75/11 **Geography**

Date — Not applicable

Duration — 2 hours 20 minutes

Total marks — 80

SECTION 1 — PHYSICAL ENVIRONMENTS — 30 marks

Attempt **EITHER** question 1 **OR** question 2

THEN attempt questions 3 to 6.

SECTION 2 — HUMAN ENVIRONMENTS — 30 marks

Attempt ALL questions.

SECTION 3 — GLOBAL ISSUES — 20 marks

Attempt any **TWO** of the following.

Question 11 — Climate change

Question 12 — Natural regions

Question 13 — Environmental hazards

Question 14 — Trade and globalisation

Question 15 — Tourism

Question 16 — Health

You will receive credit for appropriately labelled sketch maps and diagrams.

Write your answers clearly in the answer booklet provided. In the answer booklet you must clearly identify the question number you are attempting.

Use **blue** or **black** ink.

Before leaving the examination room you must give your answer booklet to the Invigilator; if you do not, you may lose all the marks for this paper.

MARKS

SECTION 1 — PHYSICAL ENVIRONMENTS — 30 marks

Attempt EITHER question 1 OR question 2

THEN questions 3 to 6

Question 1 — Glaciated landscapes

(a) Study the Ordnance Survey map extract (Item A) of the Brecon Beacons area.

Using grid references, **describe** the evidence shown on the map which suggests that this is an area of **upland glaciated scenery**. 4

(b) **Explain** the formation of a **U-shaped valley**.

You may use a diagram(s) in your answer. 4

Now attempt questions 3 to 6

MARKS

Do not attempt question 2 if you have already answered question 1

Question 2 — Rivers and valleys

(a) Study the Ordnance Survey map extract (Item A) of the Brecon Beacons area.

Describe the physical features of the Afon (River) Nedd Fechan **and** its valley between 905175 and 900092. You should use grid references in your answer. 4

(b) **Explain** the formation of an **ox-bow lake**.

You may use a diagram(s) in your answer. 4

Now attempt questions 3 to 6

MARKS

Question 3

Item Q3: Quote from a local landowner

"This area has the potential for a variety of different land uses, including farming, forestry, recreation/tourism, water storage/supply, industry and renewable energy."

Study Item Q3 and the Ordnance Survey map extract (Item A) of the Brecon Beacons area.

Choose **two** different land uses mentioned in Item Q3.

Using map evidence, **explain** how the area shown on the map extract is suitable for your chosen land uses. 5

MARKS

Question 4

Diagram Q4: Synoptic chart, 0800 hours, 10th March

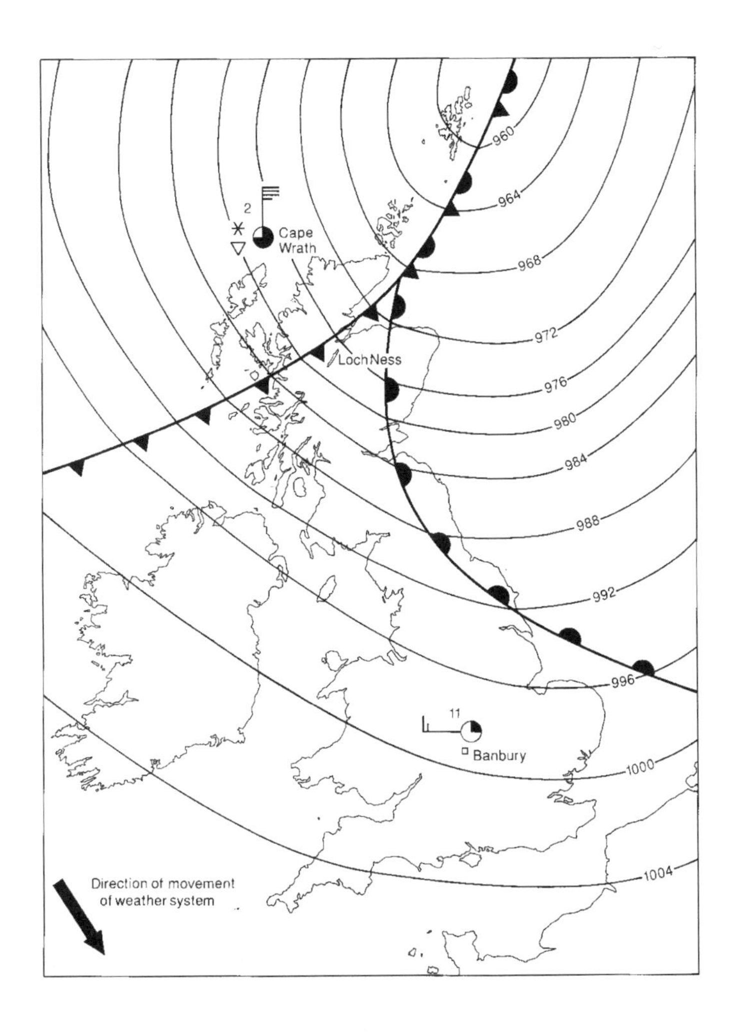

Study Diagram Q4.

(a) **Describe**, in detail, the differences in the weather between Cape Wrath and Banbury at 0800 hours on 10th March. **4**

(b) At 0800 hours on 10th March a group of secondary school students are about to set off on a walk into the mountains near Loch Ness. After seeing the weather chart in Diagram Q4, they decide to cancel their walk at the last minute.

Why might conditions have been unsuitable for their expedition? Give reasons. **5**

[Turn over

MARKS

Question 5

Diagram Q5: Air masses affecting the British Isles

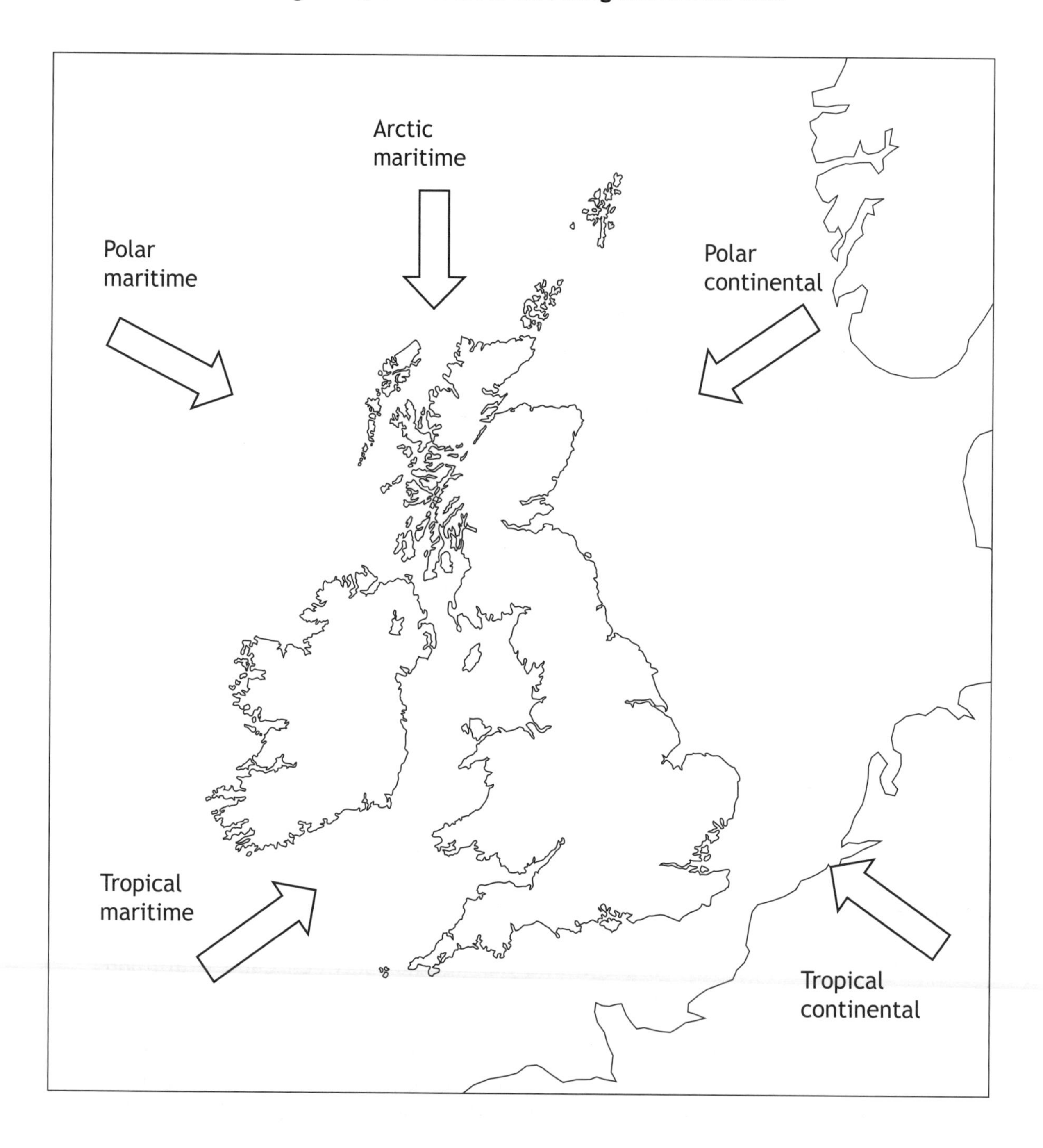

Look at Diagram Q5.

Describe how a long period with a **tropical continental** air mass **in summer** would affect the people of the British Isles. 3

MARKS

Question 6

Diagram Q6: Selected land uses

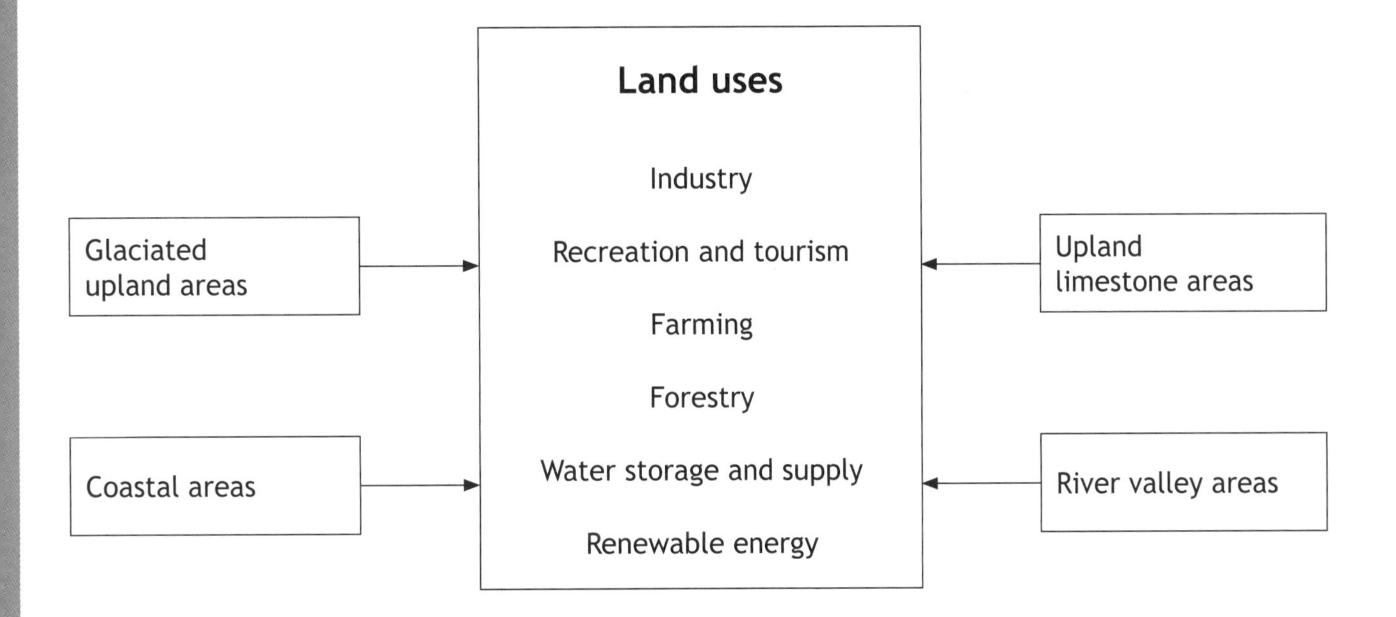

Look at Diagram Q6.

(i) For a named area you have studied, **explain**, **in detail**, ways in which **two** different land uses may be in conflict with each other.

(ii) **Suggest** possible solutions to these conflicts. 5

[Turn over

MARKS

SECTION 2 — HUMAN ENVIRONMENTS — 30 marks

Attempt ALL questions

Question 7

Diagram Q7

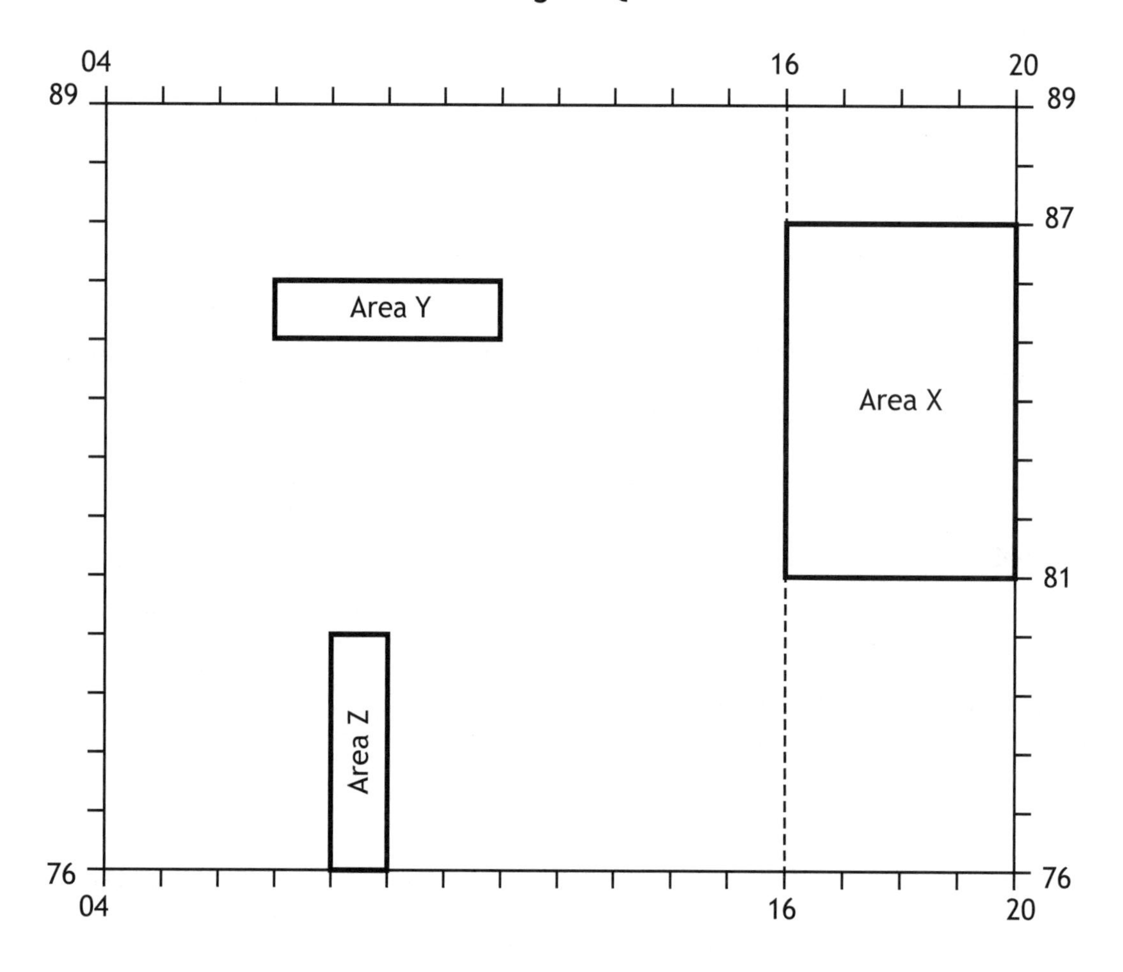

Study the Ordnance Survey map extract (Item B) of the Birmingham area and Diagram Q7 above.

(a) Give map evidence to show that part of the Central Business District (CBD) of Birmingham is found in grid square 0786. **3**

(b) Find Area X on Diagram Q7 and the map extract (Item B).

Birmingham airport, a golf course, a business park and a housing area are found in Area X on the rural/urban fringe of Birmingham. Using map evidence **explain** why such developments are found there. **5**

(c) The Russell family have three young children and are buying a house in Birmingham. They have narrowed down their search to two areas of the city – Area Y (Balsall Heath and Sparkbrook) or Area Z (Highter's Heath and Drake's Cross).

Which area, Y or Z, should they choose? Using **detailed map evidence, give reasons** to support your chosen area. **6**

MARKS

Question 8

Diagram Q8: Developments in farming

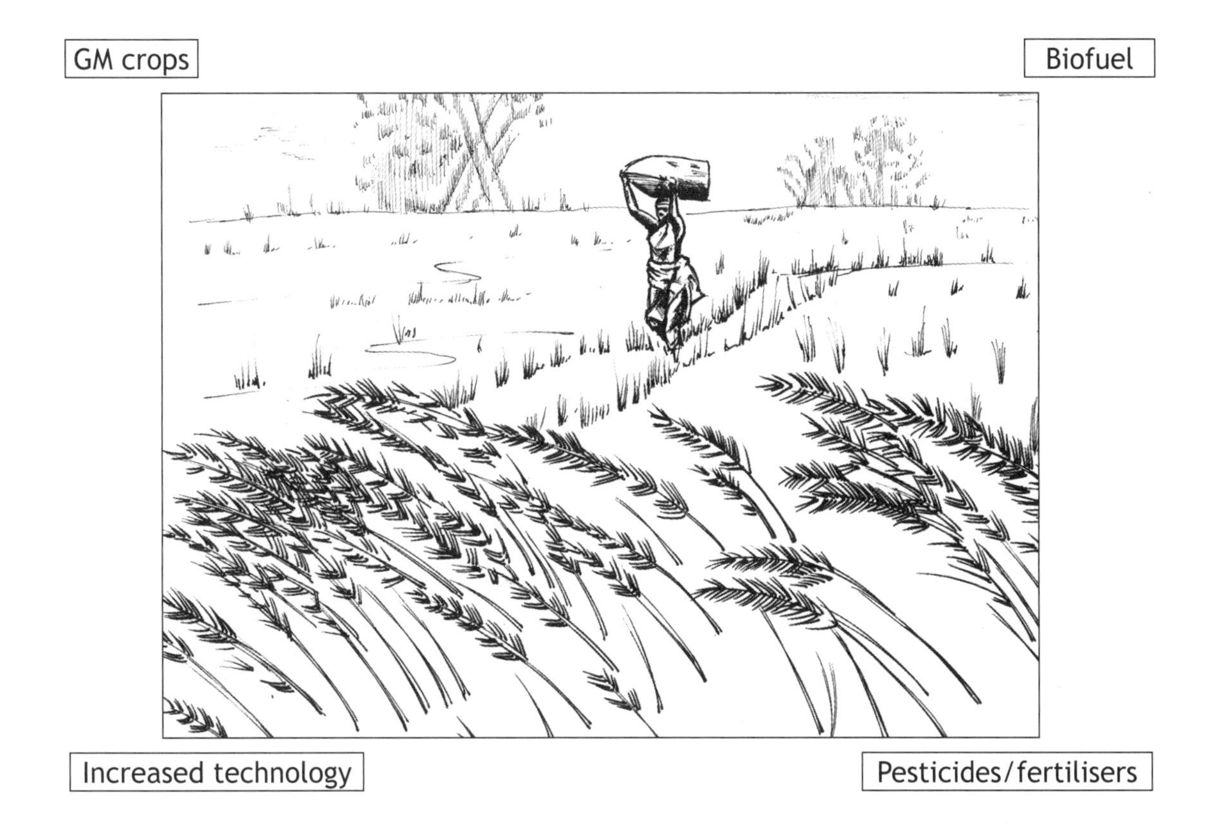

Look at Diagram Q8.

Explain how recent developments in agriculture in developing countries are helping farmers. 4

[Turn over

Question 9 **MARKS**

Diagram Q9: Demographic transition model

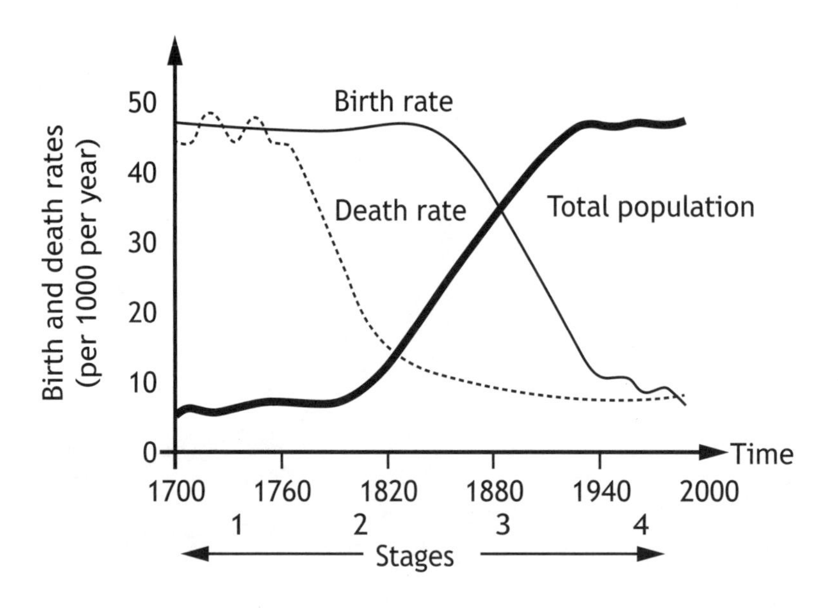

Look at Diagram Q9.

Countries such as the United Kingdom have experienced significant population change, as shown in Diagram Q9.

Explain why this population change has happened. You must refer to factors affecting birth and death rates in stages 2, 3 and 4 of Diagram Q9. 6

Question 10

Table Q10: Selected development indicators

Country	*Life expectancy (yrs)*	*Access to safe drinking water (%)*	*Literacy rate (%)*	*% of workforce employed in agriculture*
Bolivia	69	90	96	32
Chad	50	51	40	80
Finland	81	100	100	4
Mali	56	77	39	80
Netherlands	81	100	100	2
Uganda	55	79	78	40

Study Table Q10.

Choose **two** of the development indicators shown.

For the **two** that you have chosen, **explain**, **in detail**, why they are useful in helping to show a country's level of development. 6

SECTION 3 — GLOBAL ISSUES — 20 marks

Attempt any TWO questions

Question 11 — Climate change **(Page 10)**

Question 12 — Natural regions **(Page 11)**

Question 13 — Environmental hazards **(Page 12)**

Question 14 — Trade and globalisation **(Page 14)**

Question 15 — Tourism **(Page 15)**

Question 16 — Health **(Page 16)**

MARKS

Question 11: Climate change

Diagram Q11: Area of Arctic Sea ice (1979—2013)

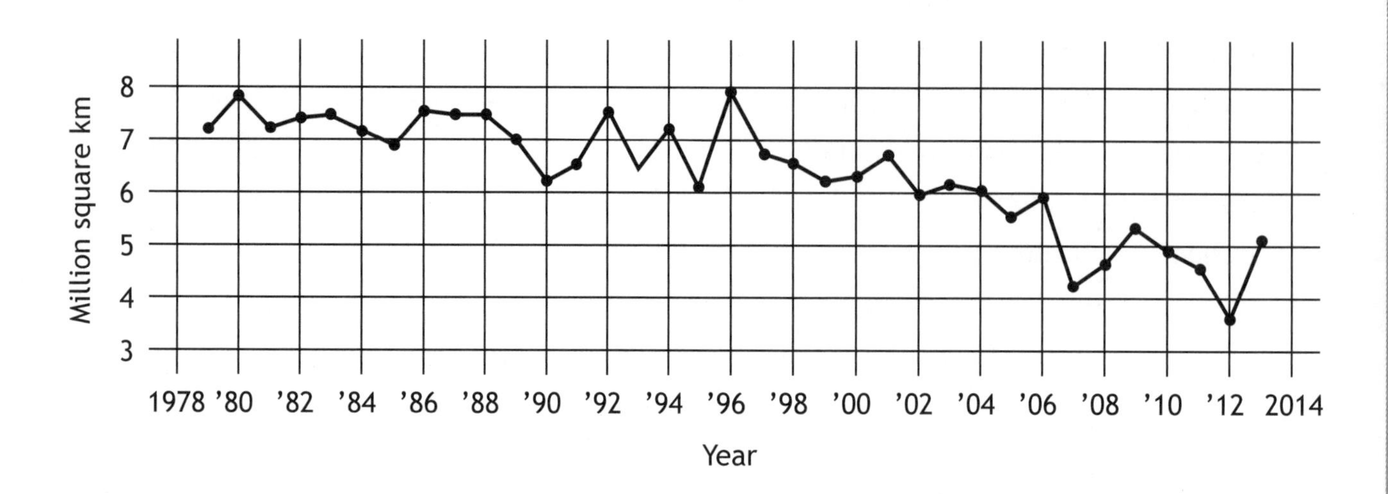

Study Diagram Q11.

(a) **Describe**, **in detail**, the changes in the area of Arctic Sea ice. **4**

(b) Melting sea ice is one effect of climate change.

Explain some other effects of climate change. **6**

MARKS

Question 12: Natural regions

Diagram Q12: Recent deforestation rates worldwide

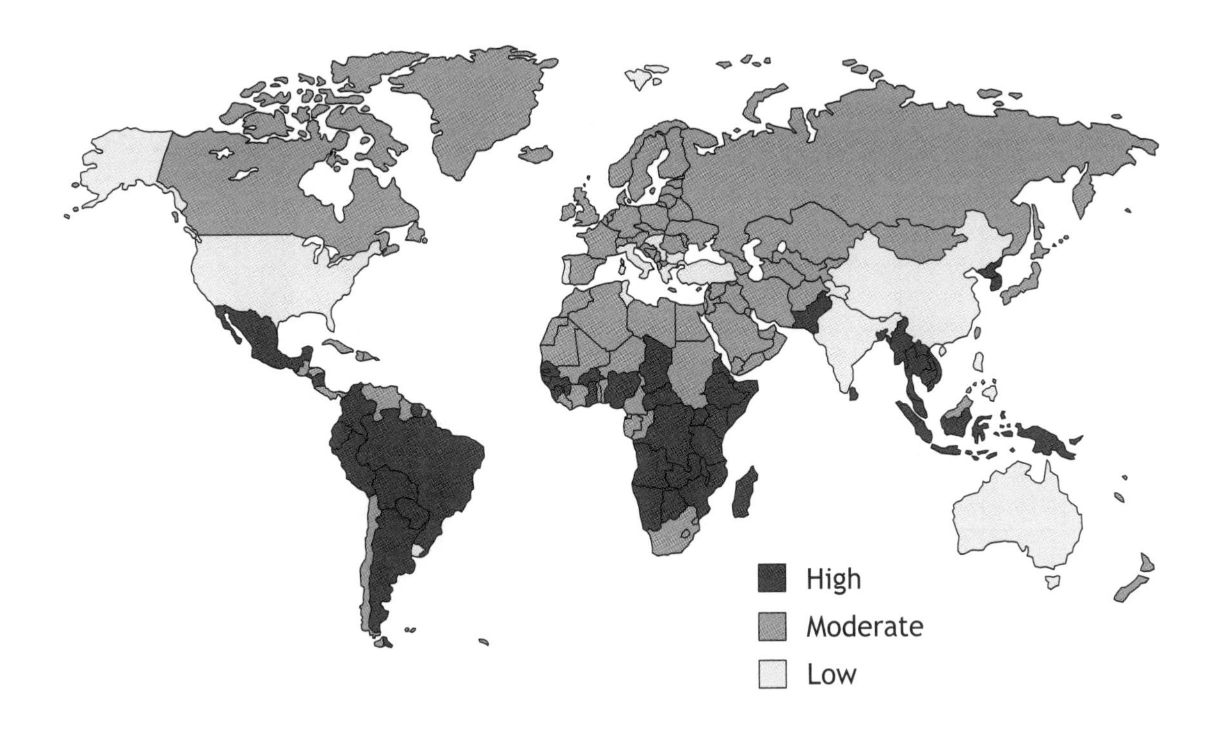

(a) Study Diagram Q12.

Describe, **in detail**, deforestation rates worldwide. 4

(b) **Explain** the management strategies which can be used to minimise the impact of human activity in the tundra. 6

[Turn over

MARKS

Question 13: Environmental hazards

Diagram Q13A: Number of volcanic eruptions per decade 1910–2010

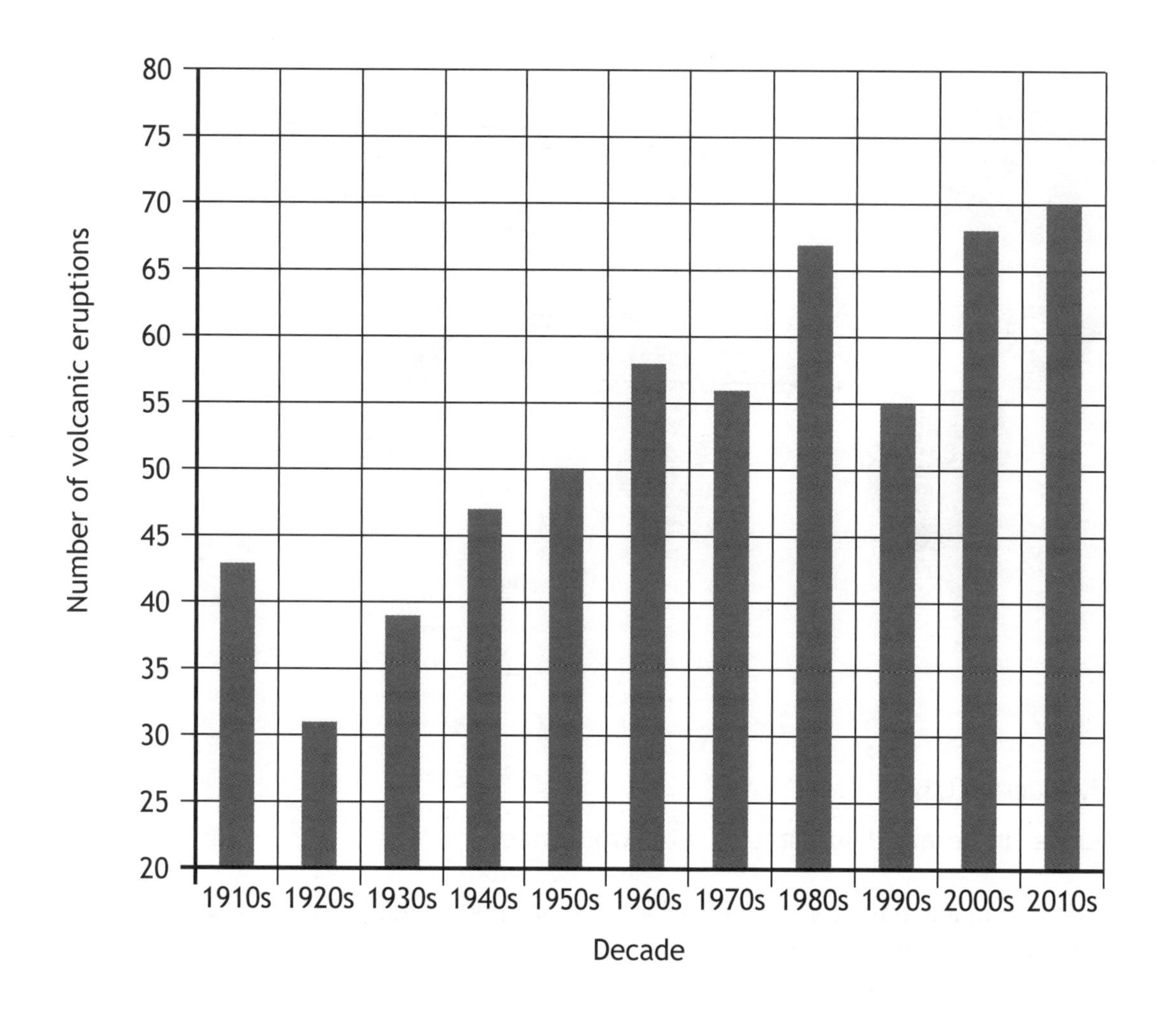

(a) Study Diagram Q13A.

Describe, **in detail**, the changes in the number of volcanic eruptions between 1910–2010. 4

MARKS

Question 13 (continued)

Item Q13B: Pico de Fogo volcano, Cape Verde

After nearly 20 years of inactivity, the Pico de Fogo awakened with a violent eruption on the 23rd of November 2014.

(b) Look at Item Q13B.

For a volcanic eruption you have studied, **explain**, **in detail**, the impacts of the eruption on people and the landscape. **6**

[Turn over

MARKS

Question 14: Trade and globalisation

Diagram Q14A: World exports by region

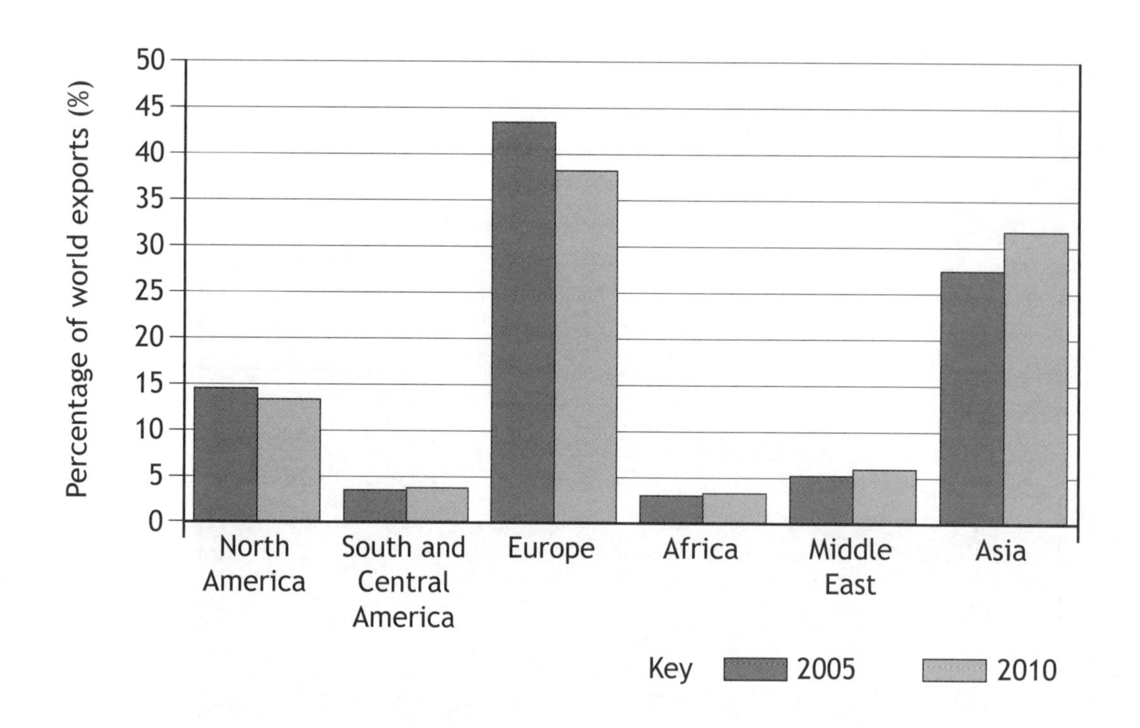

(a) Study Diagram Q14A.

Describe, in detail, the changes in world exports from 2005 to 2010. 4

Item Q14B: Collecting Fairtrade coffee beans

(b) Look at Item Q14B.

Explain how buying Fairtrade products helps people in the developing world. 6

MARKS

Question 15: Tourism

Diagram Q15A: Top ten world tourist destinations (millions of visitors per year)

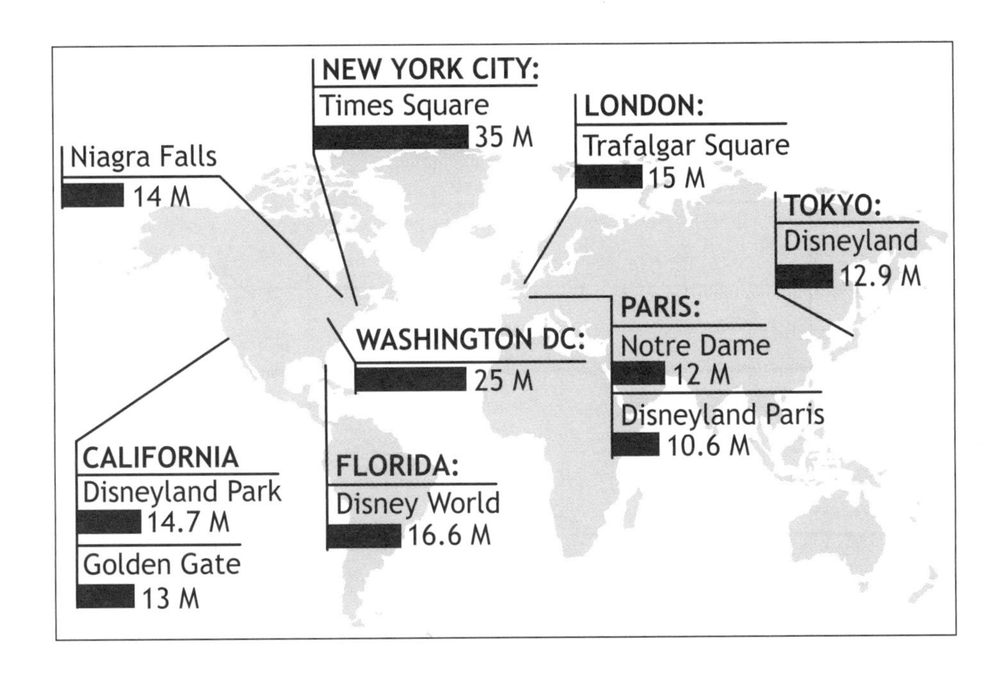

(a) Study Diagram Q15A.

Describe, **in detail**, the distribution of the top ten world tourist destinations. **4**

Item Q15B — Mass tourism on an Italian beach

(b) Look at Item Q15B.

Describe the effects of mass tourism on people and the environment. **6**

MARKS

Question 16: Health

Diagram Q16: Ebola cases in selected African countries April–Oct 2014

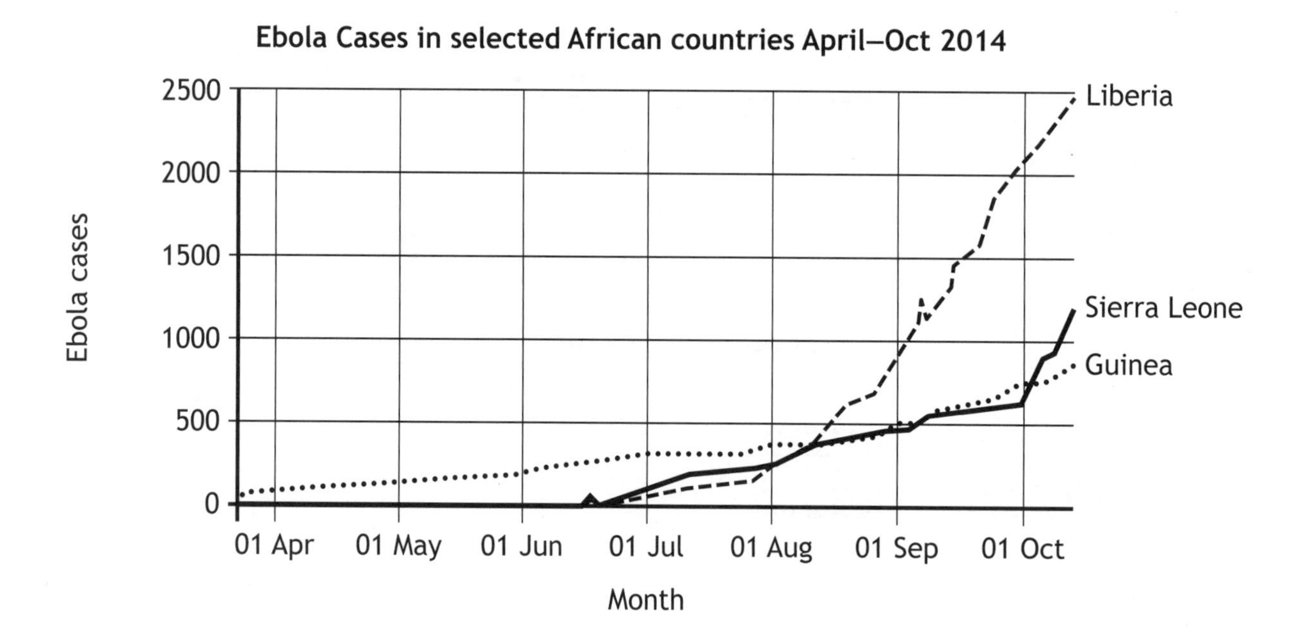

(a) Study Diagram Q16.

Describe, **in detail**, the changes in Ebola cases in the **three** named African countries. 4

(b) **Explain** the causes of either heart disease **or** cancer **or** asthma. 6

[END OF SPECIMEN QUESTION PAPER]

S833/75/11

Geography
Ordnance Survey Map
Item A

Date — Not applicable

Duration — 2 hours 20 minutes

The colours used in the printing of these map extracts are indicated in the four little boxes at the top of the map extract. Each box should contain a colour; if any does not, the map is incomplete and should be returned to the Invigilator.

Extract No 2213/160

Four colours should appear above; if not then please return to the invigilator.

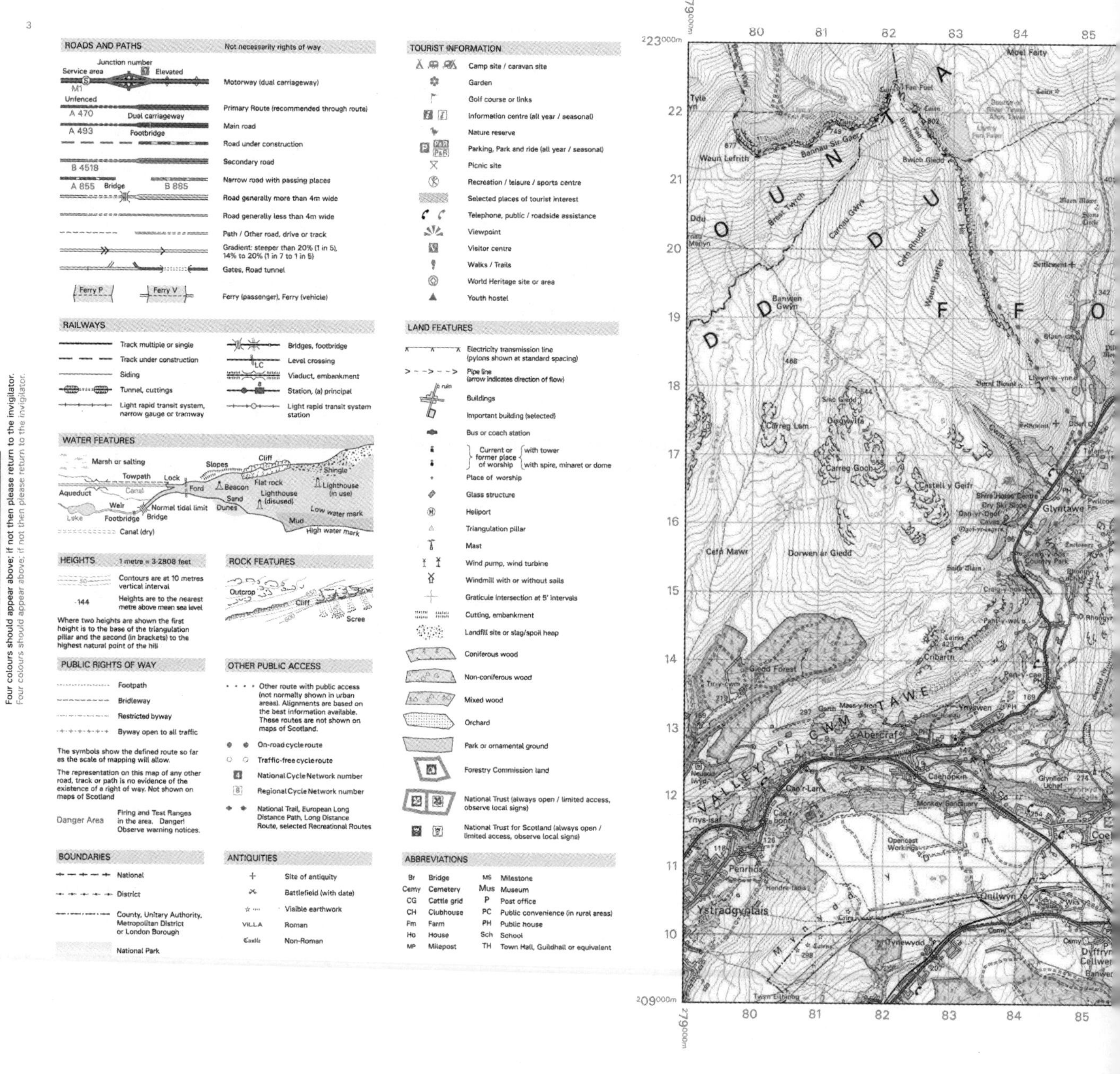

1:50 000 Scale
Landranger Series

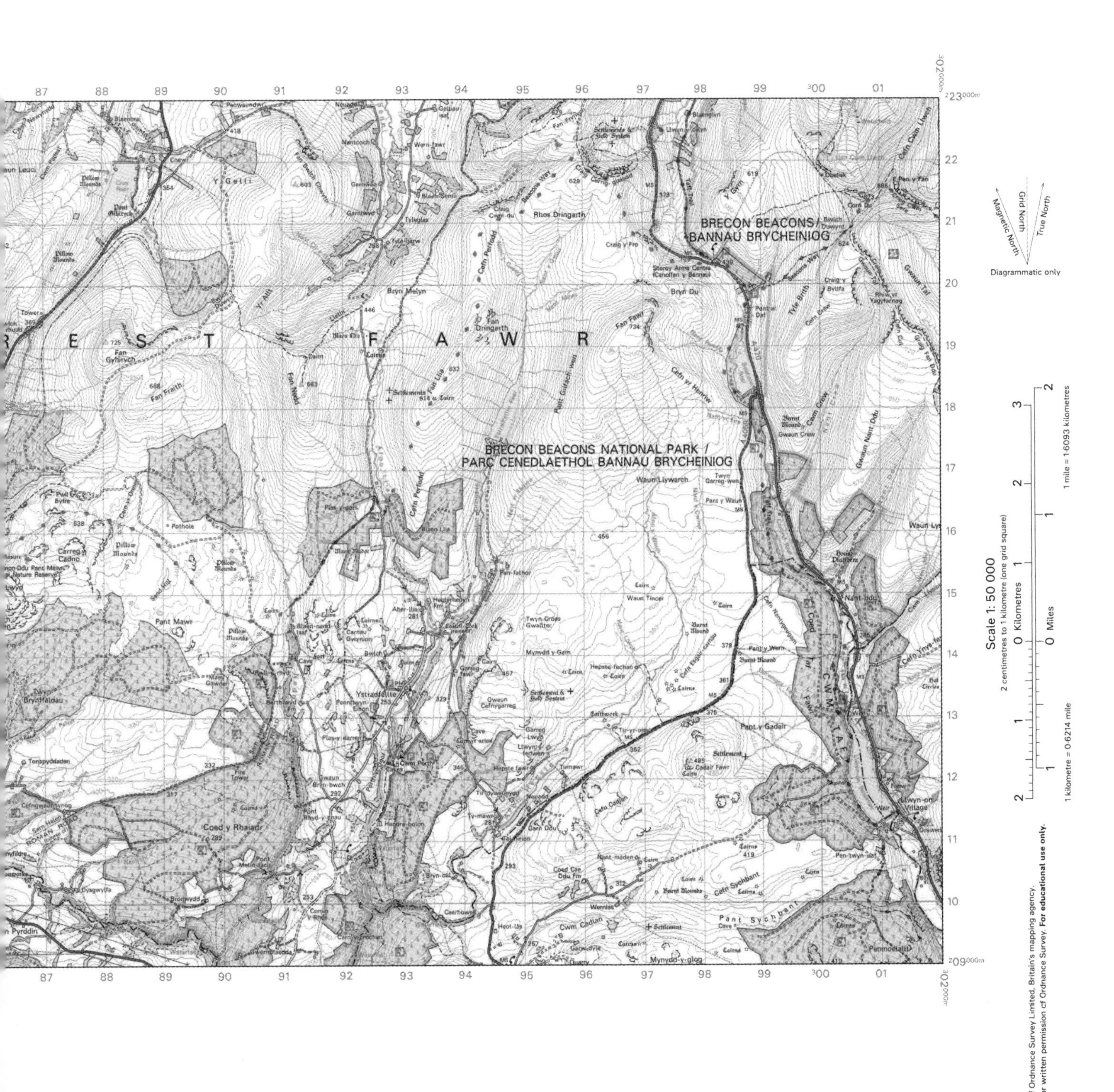
BRECON BEACONS NATIONAL PARK /
PARC CENEDLAETHOL BANNAU BRYCHEINIOG
BRECON BEACONS/
BANNAU BRYCHEINIOG
E S T F A W R
Ystradfellte
Coed y Rhaiadr
Fan Gyhirych
Fan Fraith
Fan Nedd
Fan Llia
Fan Dringarth
Fan Fawr
Bryn Melyn
Rhos Dringarth
Bryn Du
Storey Arms Centre
Cefn yr Henriw
Waun Llywarch
Mynydd y Garn
Twyn Bryn-y-glog
Pant Mawr
Pant y Gadair
Cefn Sychbant
Coed Taf Fawr
CWM TAF
Nant-ddu
Llwyn-on Village
Penmoelallt
Mynydd-y-glog
87 88 89 90 91 92 93 94 95 96 97 98 99 300 01
23 22 21 20 19 18 17 16 15 14 13 12 11 10 09
Grid North
True North
Magnetic North
Diagrammatic only
Scale 1: 50 000
2 centimetres to 1 kilometre (one grid square)
Kilometres
Miles
1 mile = 1·6093 kilometres
1 kilometre = 0·6214 mile

[BLANK PAGE]

DO NOT WRITE ON THIS PAGE

S833/75/11

Geography
Ordnance Survey Map
Item B

Date — Not applicable

Duration — 2 hours 20 minutes

The colours used in the printing of these map extracts are indicated in the four little boxes at the top of the map extract. Each box should contain a colour; if any does not, the map is incomplete and should be returned to the Invigilator.

Extract No 2142/139

Four colours should appear above; if not then please return to the invigilator.

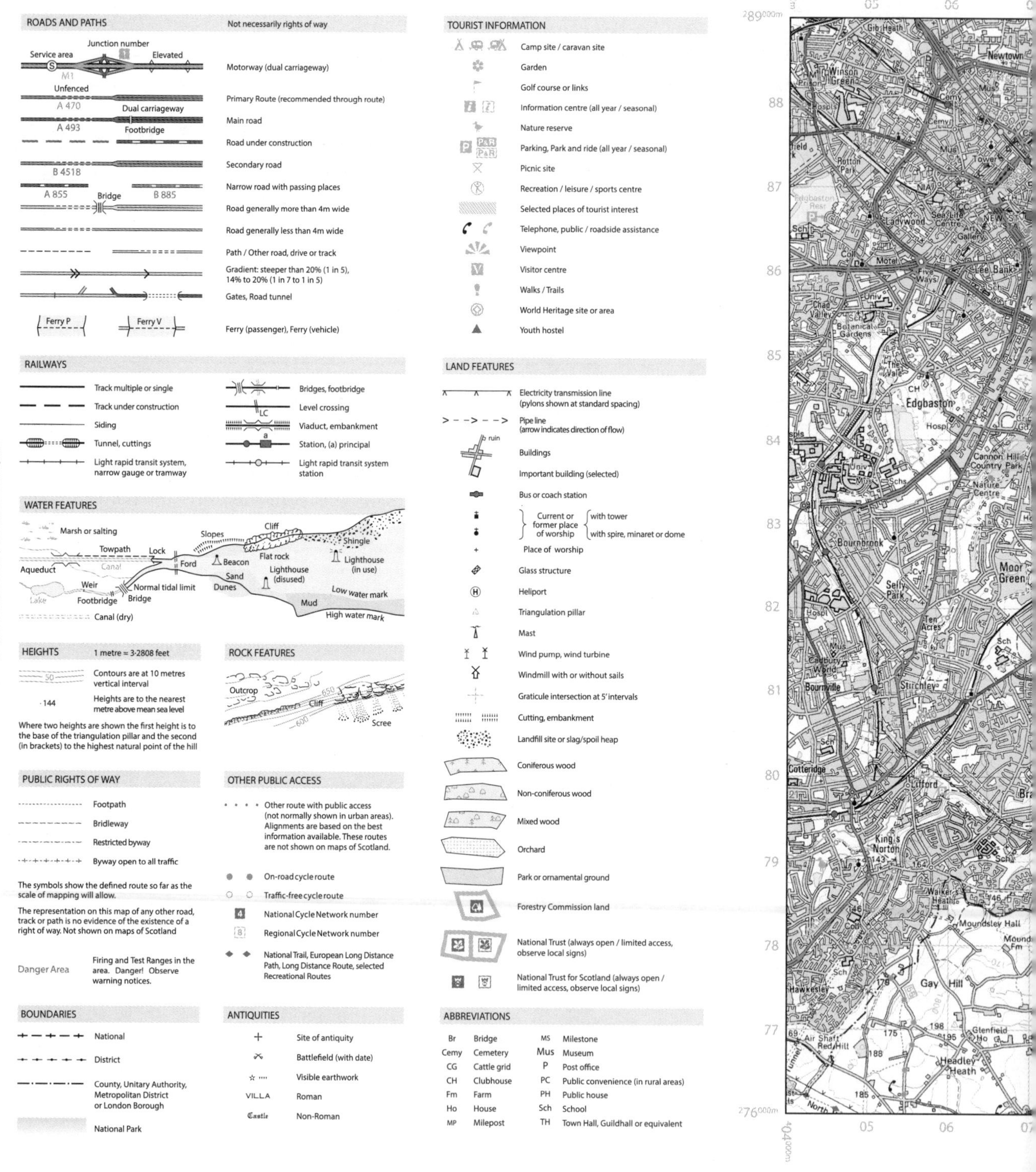

1:50 000 Scale
Landranger Series

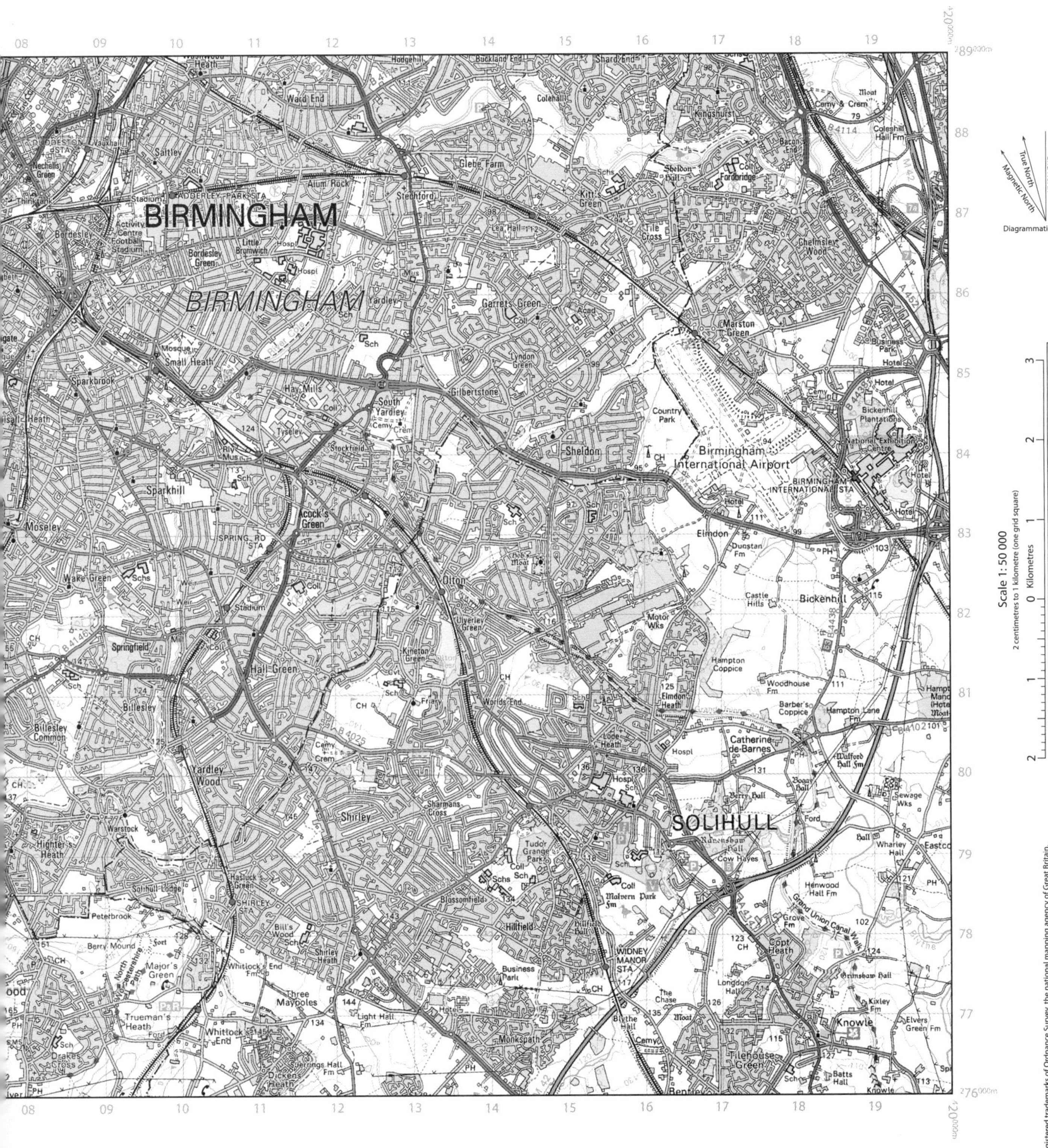
BIRMINGHAM
SOLIHULL
Birmingham International Airport
BIRMINGHAM INTERNATIONAL STA
National Exhibition Centre
Saltley
Stechford
Sheldon
Small Heath
Sparkbrook
Sparkhill
Acock's Green
Moseley
Hall Green
Yardley Wood
Shirley
Olton
Elmdon
Bickenhill
Catherine-de-Barnes
Kingshurst
Marston Green
Chelmsley Wood
Knowle
Monkspath
Hillfield
Grid North
True North
Magnetic North
Diagrammatic only
Scale 1: 50 000
2 centimetres to 1 kilometre (one grid square)
Kilometres
Miles
1 mile = 1·6093 kilometres
1 kilometre = 0·6214 mile

[BLANK PAGE]

DO NOT WRITE ON THIS PAGE

NATIONAL 5

2018

X833/75/11 **Geography**

TUESDAY, 1 MAY

1:00 PM – 3:20 PM

Total marks — 80

SECTION 1 — PHYSICAL ENVIRONMENTS — 30 marks

Attempt **EITHER** question 1 **OR** question 2.

THEN attempt questions 3 to 7.

SECTION 2 — HUMAN ENVIRONMENTS — 30 marks

Attempt ALL questions.

SECTION 3 — GLOBAL ISSUES — 20 marks

Attempt any **TWO** of the following.

Question 13 — Climate change

Question 14 — Natural regions

Question 15 — Environmental hazards

Question 16 — Trade and globalisation

Question 17 — Tourism

Question 18 — Health

You will receive credit for appropriately labelled sketch maps and diagrams.

Write your answers clearly in the answer booklet provided. In the answer booklet you must clearly identify the question number you are attempting.

Use **blue** or **black** ink.

Before leaving the examination room you must give your answer booklet to the Invigilator; if you do not, you may lose all the marks for this paper.

MARKS

SECTION 1 — PHYSICAL ENVIRONMENTS — 30 marks

Attempt EITHER question 1 OR question 2

THEN questions 3 to 7

Question 1 — Coastal landscapes

(a) Study the Ordnance Survey map extract (Item A) of the Strathy area.

Match these grid references with the correct coastal features.

Grid references: **827694, 812681 and 843662.**

Choose from features: **cliff; stack; sand spit; arch.** 3

(b) Explain the formation of a sand spit. You may use a diagram(s) in your answer. 4

Now attempt questions 3 to 7

MARKS

Do not attempt question 2 if you have already answered question 1

Question 2 — Rivers and their valleys

(a) Study the Ordnance Survey map extract (Item A) of the Strathy area.

Match these grid references with the correct river features.

Grid references: **893618, 883627 and 895589.**

Choose from features: **v-shaped valley; flood plain; meander; ox-bow lake.** 3

(b) Explain the formation of a meander. You may use a diagram(s) in your answer. 4

Now attempt questions 3 to 7

[Turn over

Question 3

Diagram Q3A: Cross-section GR 858624 to GR 910590

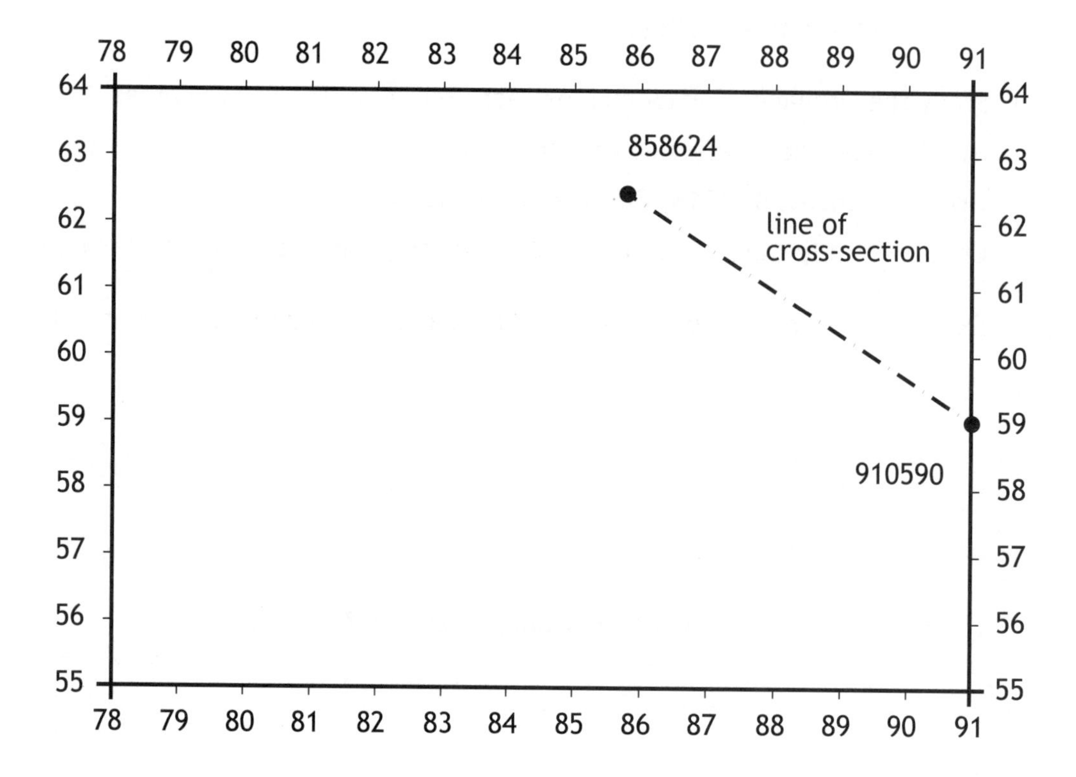

MARKS

Question 3 (continued)

Diagram Q3B: Cross-section

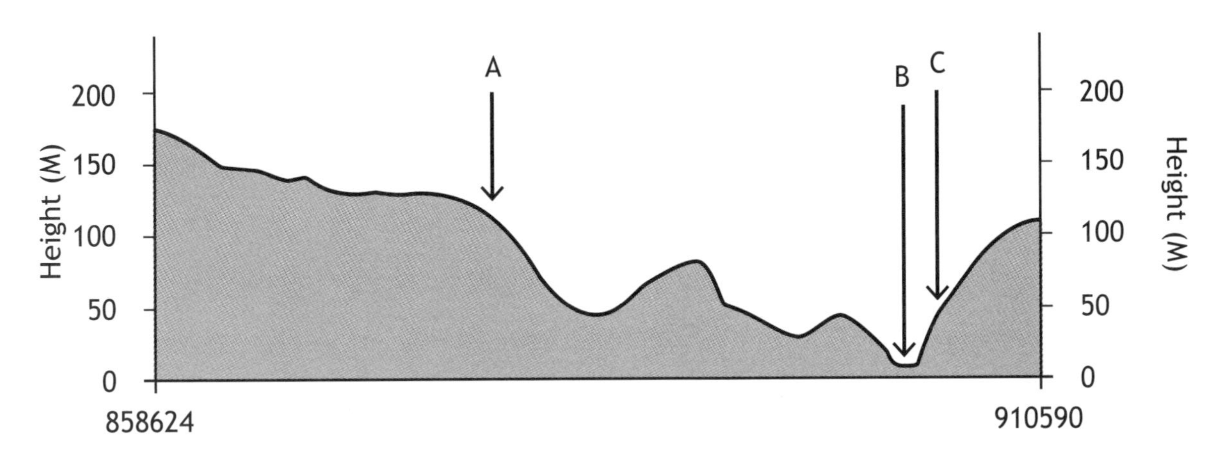

Study the Ordnance Survey map extract (Item A) of the Strathy area and Diagrams Q3A and Q3B.

Match the letters A–C with the correct features. **3**

Choose from the features below.

Halladale River; track; forestry; electricity transmission lines.

[Turn over

MARKS

Question 4

Diagram Q4: Quote from local council official

"This area has the potential for a variety of different land uses including
- farming
- forestry
- recreation/tourism
- water storage/supply
- industry
- renewable energy."

Study Diagram Q4 and the Ordnance Survey map extract (Item A) of the Strathy area.

Choose **two** different land uses listed in Diagram Q4.

Using map evidence, **explain** how the area shown on the map extract is suitable for your two chosen land uses. 5

MARKS

Question 5

Diagram Q5: Selected land uses

Look at Diagram Q5.

Choose **one** landscape type from Diagram Q5.

For a named area you have studied, **explain in detail** ways in which land use conflicts may be managed. 6

[Turn over

MARKS

Question 6

Diagram Q6: Average annual UK temperatures

Look at Diagram Q6.

Explain the factors which affect average temperatures in the UK. 4

MARKS

Question 7

Diagram Q7: Synoptic chart for Monday, 16 April 2016 at 8am

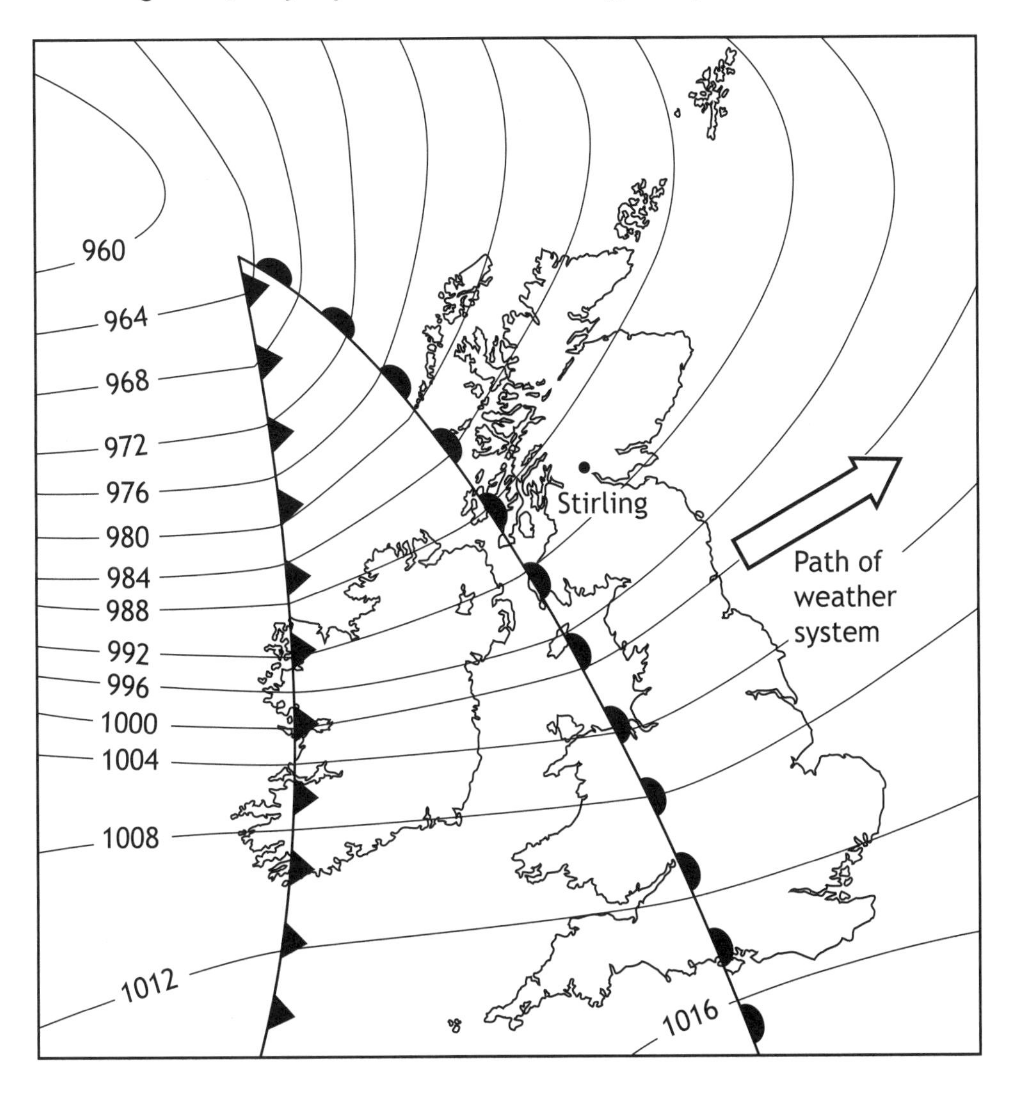

Study Diagram Q7 above.

Explain the changes that will take place in the weather in Stirling over the next 24 hours. **5**

[Turn over

MARKS

SECTION 2 — HUMAN ENVIRONMENTS — 30 marks

Attempt ALL questions

Question 8

Study the Ordnance Survey map extract (Item B) of the Oxford area.

Measure the three distances (A, B and C) between the places shown in the table.

Match your answers for A, B and C with the distances given below.

A	From the public telephone in Henwood (4702) to the school near Rose Hill (5303)
B	From Forest Farm (5410) to the church in Stanton St John (5709)
C	From Waterperry Gardens (6206) to the College (5502)

Choose from the following distances:

8·25 km   3·75 km   6·25 km   12·5 km   3

MARKS

Question 9

Diagram Q9

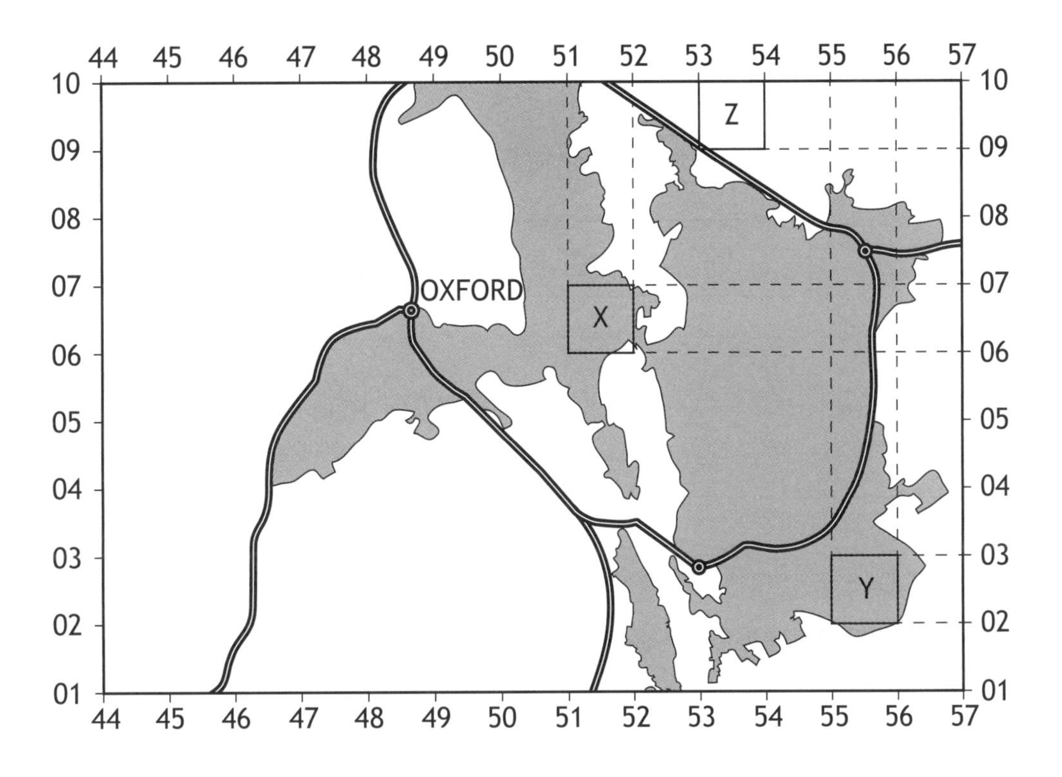

Study Diagram Q9 and the Ordnance Survey map extract (Item B) of the Oxford area.

(a) Give map evidence to explain why Area X is the CBD **and** Area Y is the suburbs. 4

(b) There is a proposal to build a new supermarket in grid square 5309 (Area Z).

Give the advantages **and** disadvantages of Area Z for this development.

You must use map evidence in your answer. 5

[Turn over

MARKS

Question 10

Diagram Q10: Inner city problems

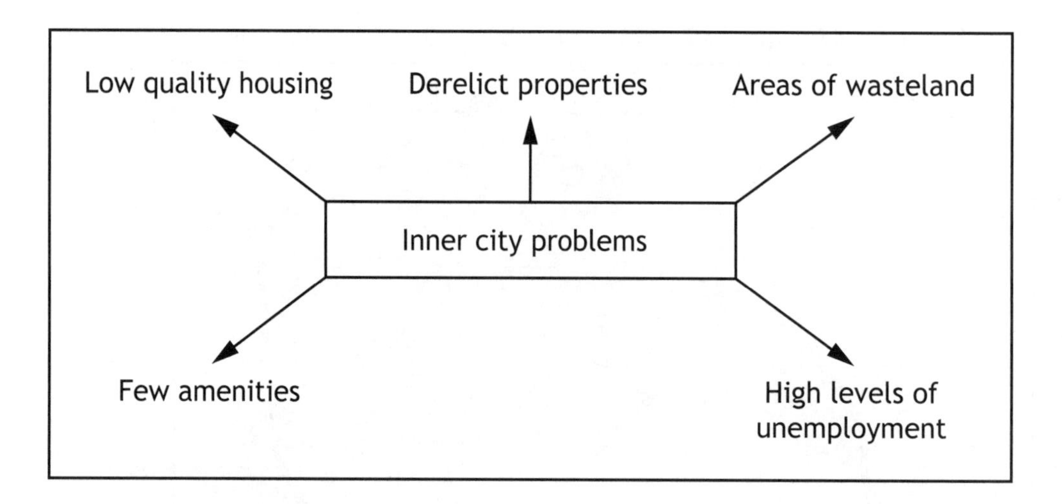

Look at Diagram Q10.

Referring to a **developed** world city you have studied, give reasons for recent changes which have taken place in the inner city. 6

MARKS

Question 11

Diagram Q11: Changes in farming in developing countries

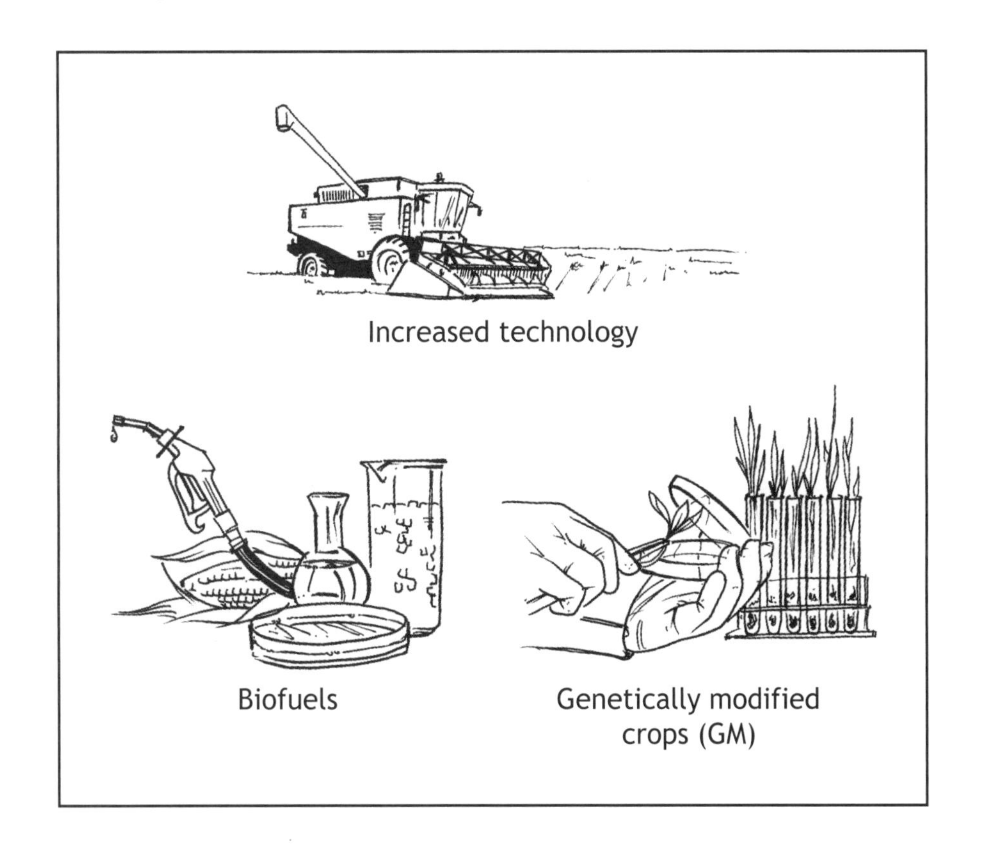

Look at Diagram Q11.

Describe, in detail, the effects of recent changes in farming on people **and** the landscape in **developing** countries.

You must mention at least **two** recent changes. 4

[Turn over

MARKS

Question 12

Diagram Q12A: Gross national income 2015 (total per country)

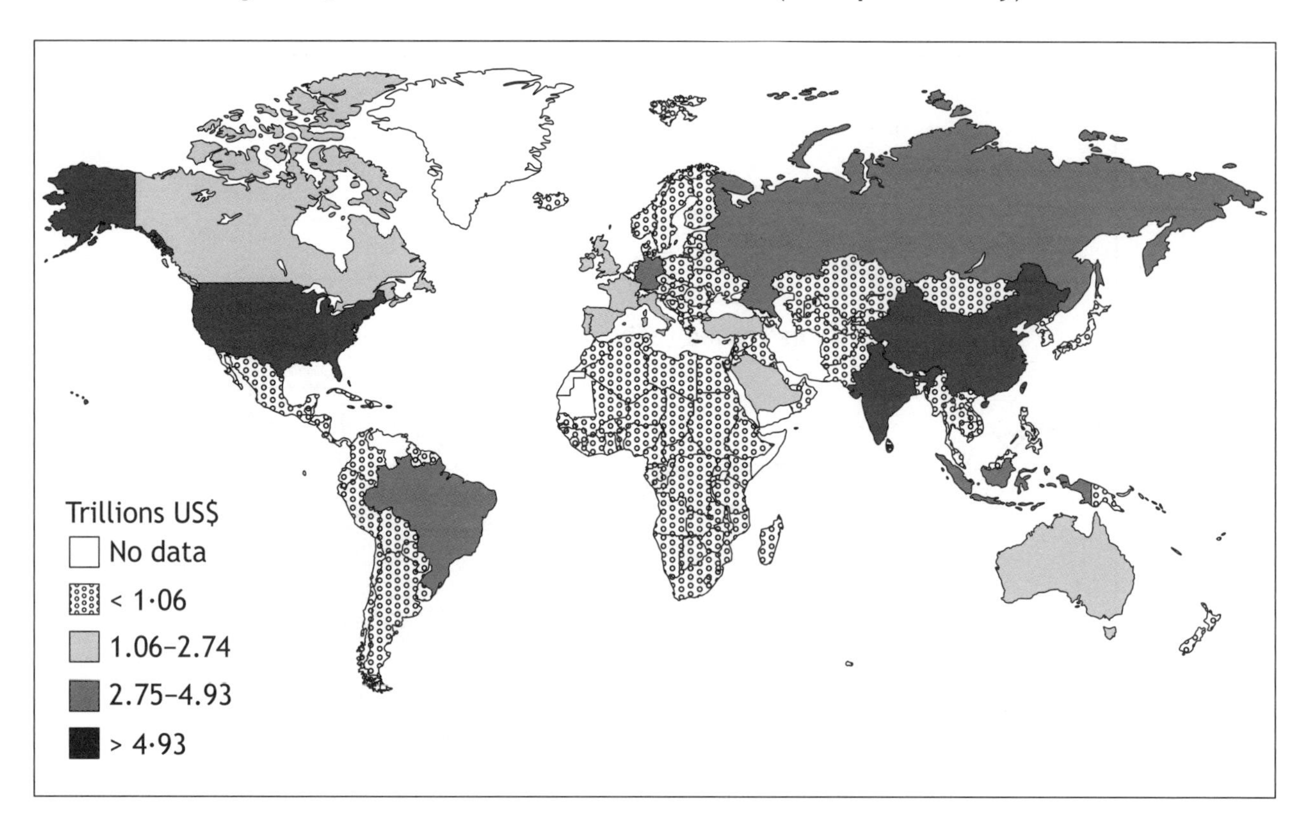

(a) Study Diagram Q12A.

Describe, in detail, the different Gross National Incomes in 2015 worldwide. 4

Diagram Q12B: Selected indicators of development

Social indicator	Economic indicator
Number of people per doctor	% of people working in agriculture
% of people who can read and write	Average income per person per year
Number of births per 1,000 women per year	Gross Domestic Product (GDP) per year

(b) Look at Diagram Q12B.

Choose **one** social and **one** economic indicator of development shown in the table.

Explain how your two chosen indicators show the level of development in a country. 4

SECTION 3 — GLOBAL ISSUES — 20 marks

Attempt any TWO questions

Question 13 — Climate change **(Page 16)**
Question 14 — Natural regions **(Page 18)**
Question 15 — Environmental hazards **(Page 20)**
Question 16 — Trade and globalisation **(Page 22)**
Question 17 — Tourism **(Page 23)**
Question 18 — Health **(Page 24)**

[Turn over

MARKS

Question 13: Climate change

Diagram Q13A: Worldwide greenhouse gas emissions (1990 to 2010)

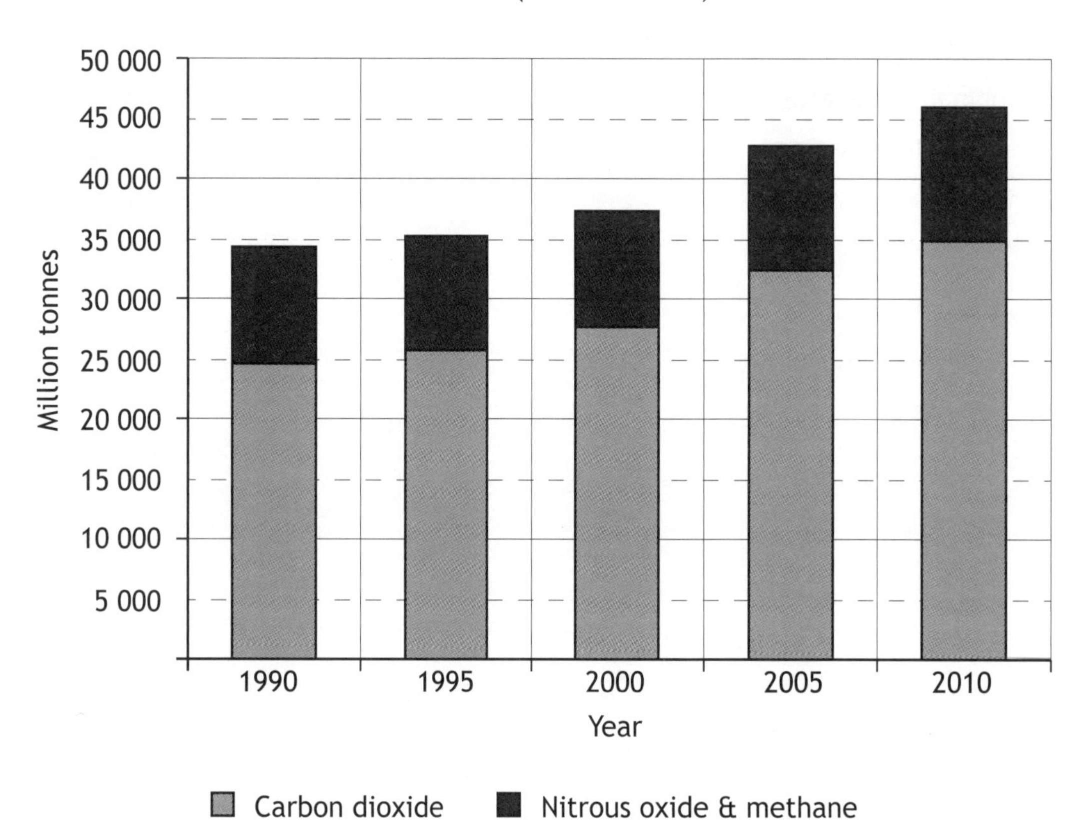

(a) Study Diagram Q13A.

Describe, in detail, the changes in greenhouse gas emissions between 1990 and 2010. 4

MARKS

Question 13 (continued)

Diagram Q13B: Online newspaper report

(b) Look at Diagram Q13B.

Explain the physical **and** human causes of climate change. 6

[Turn over

MARKS

Question 14: Natural regions

Diagram Q14A: Changes in deforestation and world population: 1900 to 2010

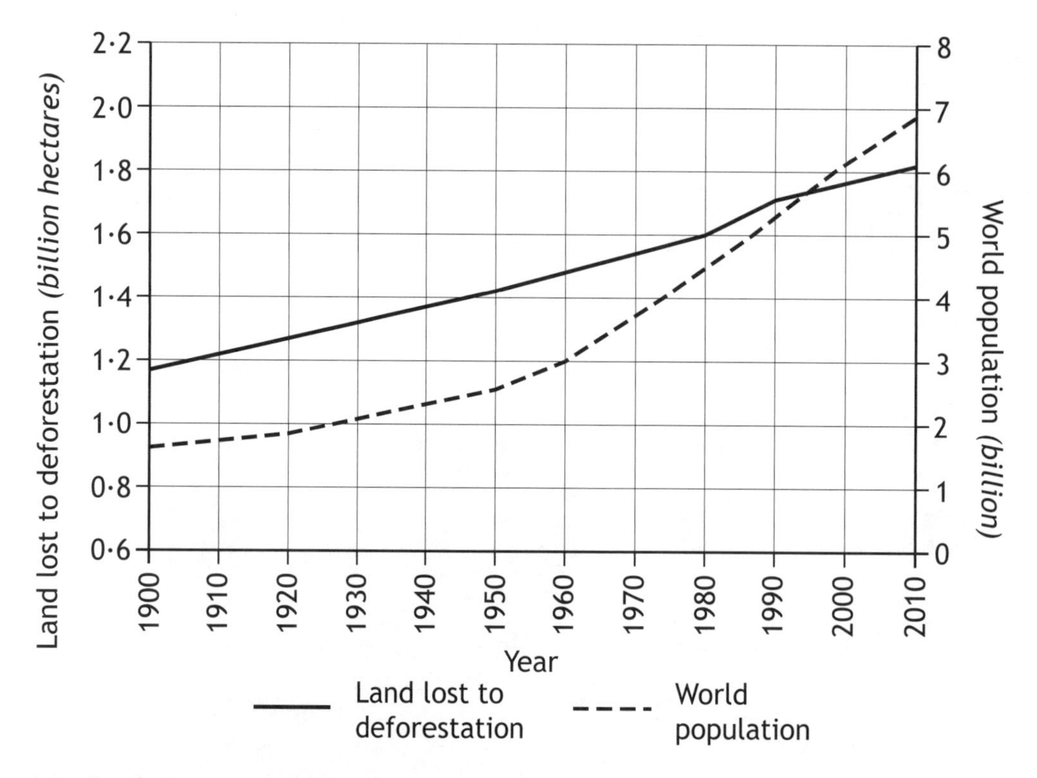

(a) Study Diagram Q14A.

Describe, in detail, changes in deforestation **and** population between 1900 and 2010. 4

MARKS

Question 14 (continued)

Diagram Q14B: Examples of vegetation adaptations in equatorial rainforests

Diagram Q14C: Examples of vegetation adaptations in the tundra

(b) Look at Diagrams Q14B and Q14C.

Explain ways in which vegetation has adapted to the environment in **either** the rainforest **or** the tundra. 6

[Turn over

MARKS

Question 15: Environmental hazards

Diagram Q15A:
Estimated damage (US$ millions) caused by natural disasters 1990 to 2012

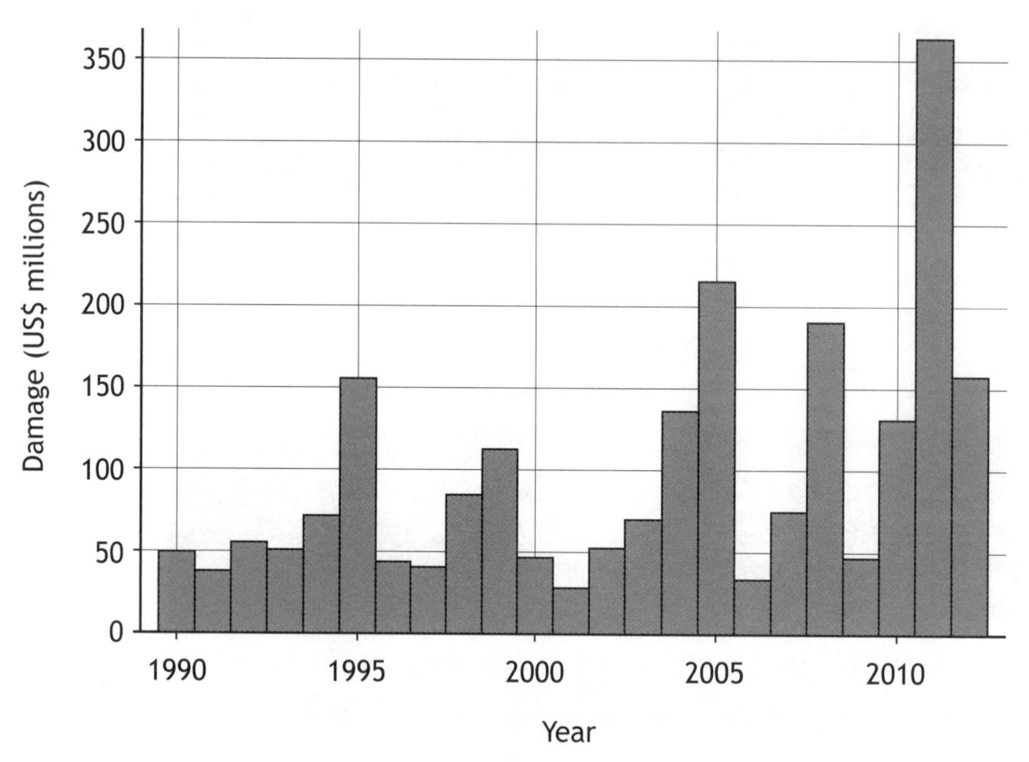

(a) Study Diagram Q15A.

Describe, in detail, the changes in estimated damage caused by natural disasters from 1990 to 2012. 4

MARKS

Question 15 (continued)

Diagram Q15B: Natural hazards in the news

(b) Look at Diagram Q15B.

For the volcano(es) that you have studied, **explain in detail** the strategies used to prepare for and reduce the effects of an eruption. **6**

[Turn over

MARKS

Question 16: Trade and globalisation

Diagram Q16: Number of Fair Trade employees 2013 to 2014

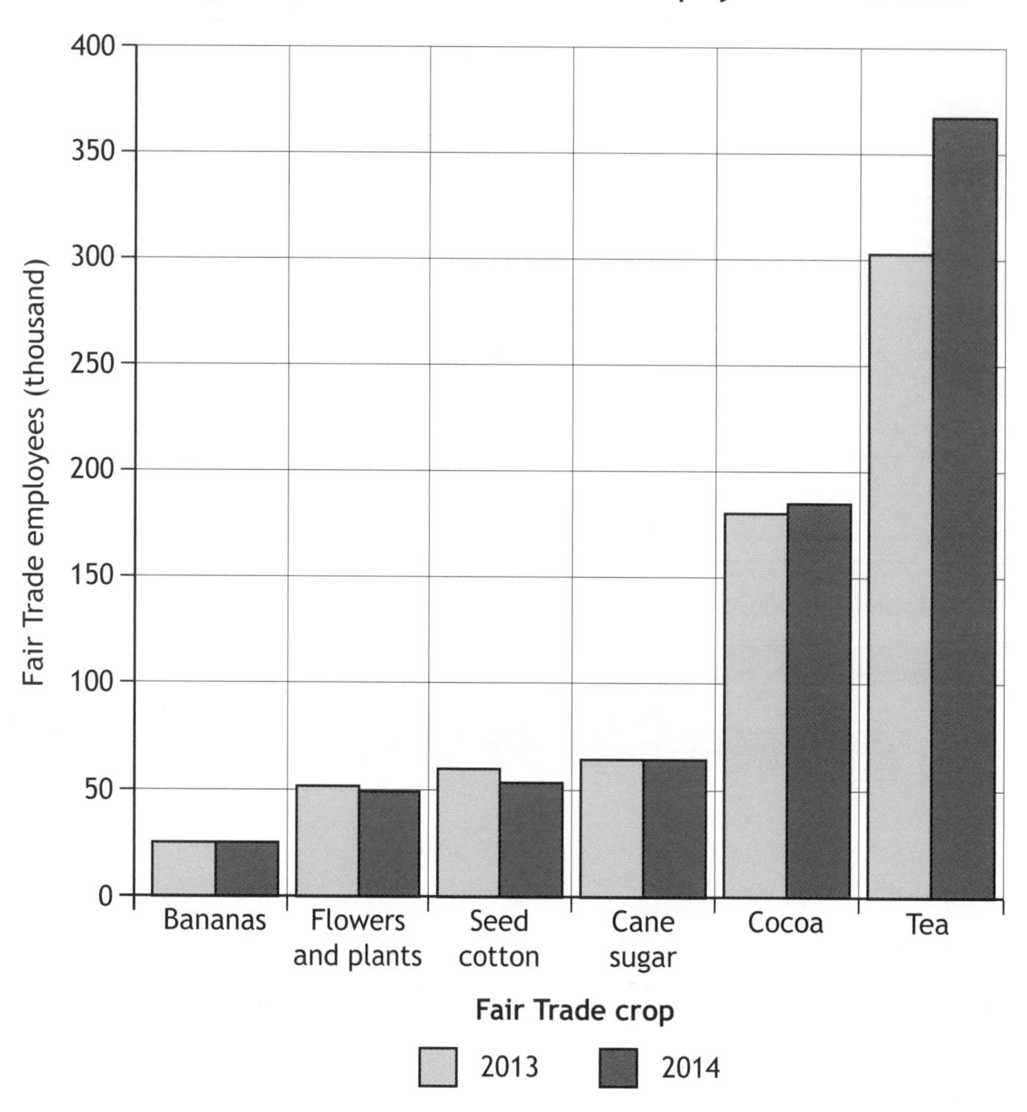

(a) Study Diagram Q16.

Describe, in detail, the changes in the number of Fair Trade employees from 2013 to 2014. 4

(b) Referring to a country or countries you have studied, **explain** how Fair Trade can help people. 6

MARKS

Question 17: Tourism

Diagram Q17: World heritage sites facing threat from tourism

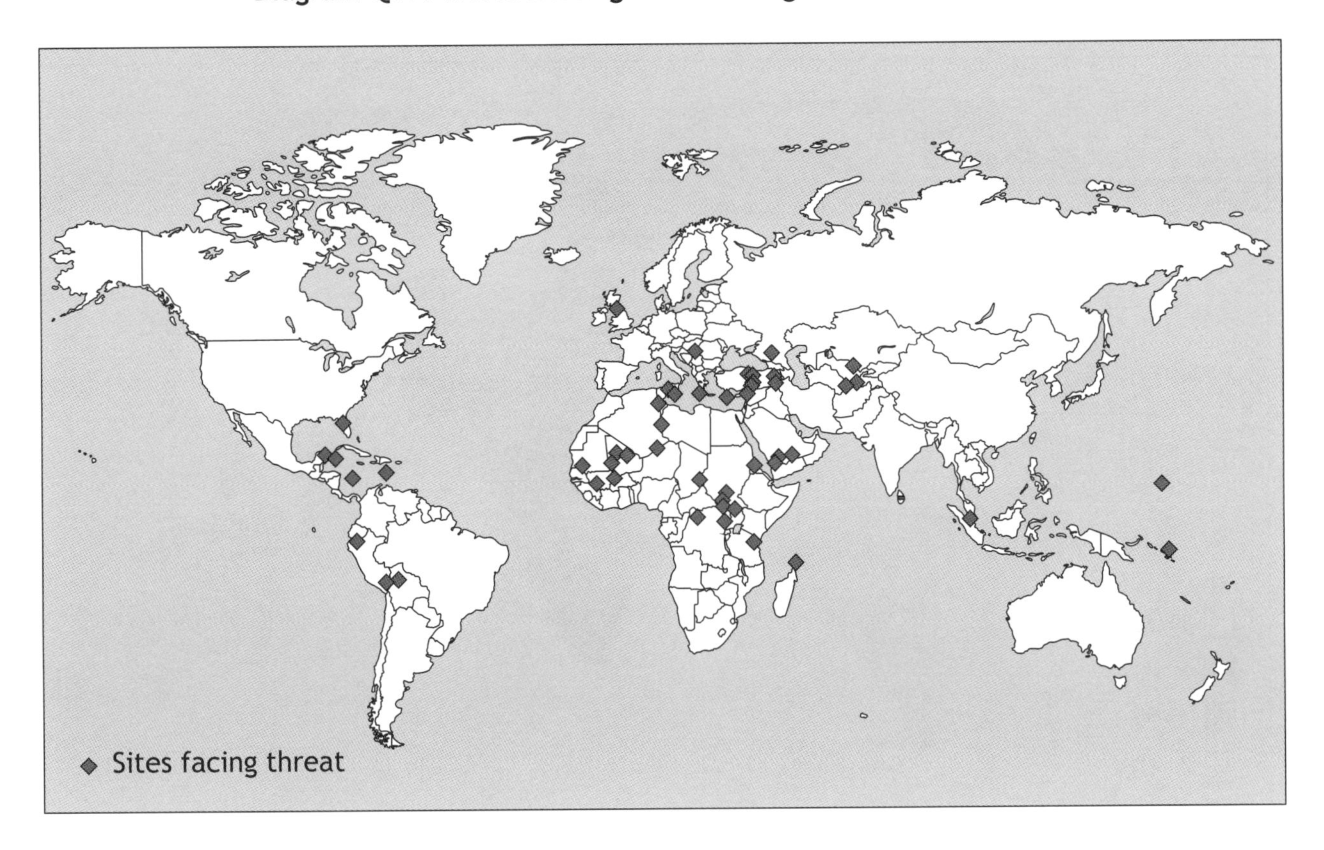

(a) Study Diagram Q17.

Describe, in detail, the location of world heritage sites facing threat from tourism. 4

(b) For named areas that you have studied, **describe** ways eco-tourism can be managed. 6

[Turn over for next question

MARKS

Question Q18: Health

Diagram Q18: Adult HIV infection rate by country 2013

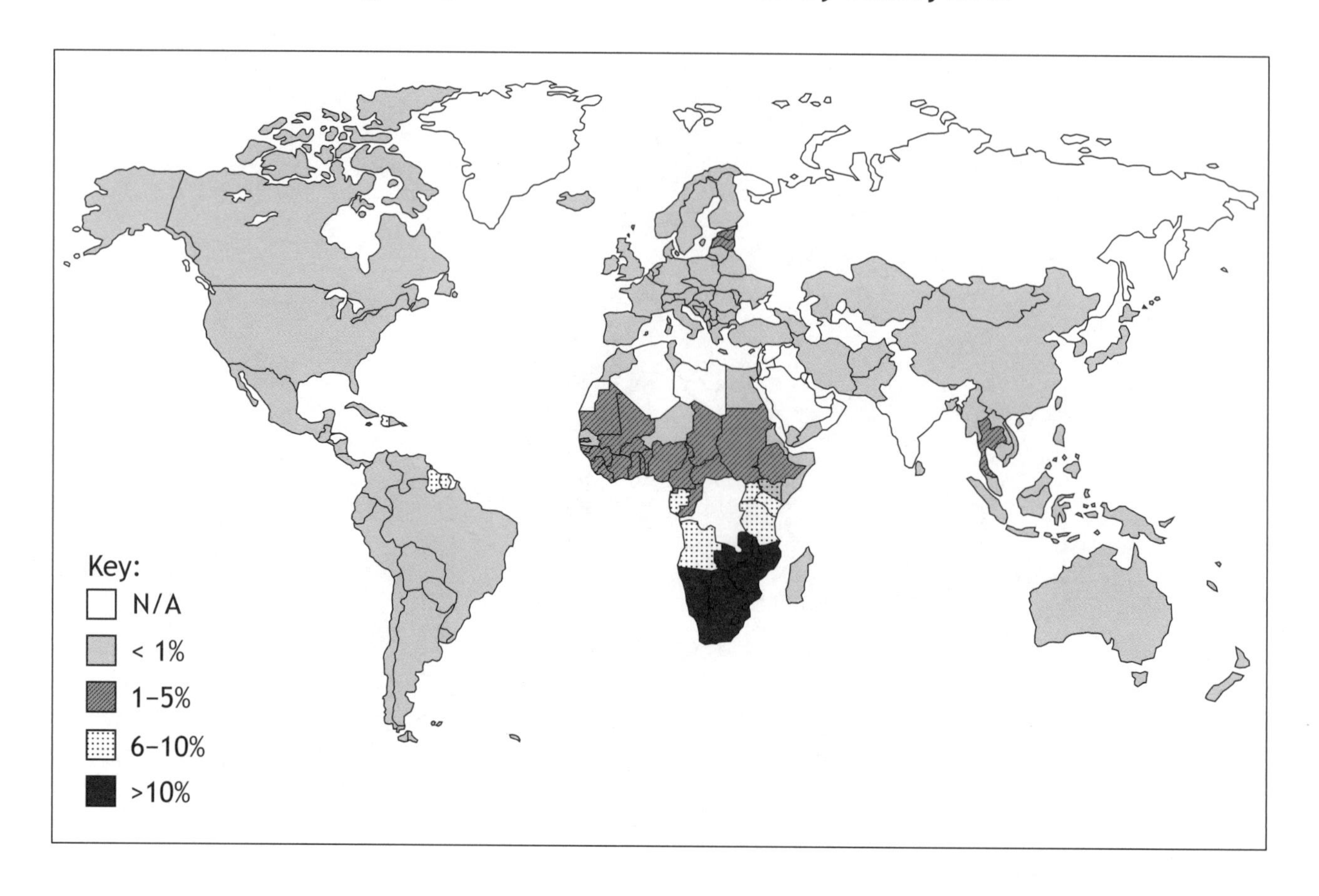

(a) Study Diagram Q18.

Describe, in detail, the global distribution of HIV/AIDS infection amongst adults. 4

(b) **Explain** the effects of HIV/AIDs on the populations of **developing** countries. 6

[END OF QUESTION PAPER]

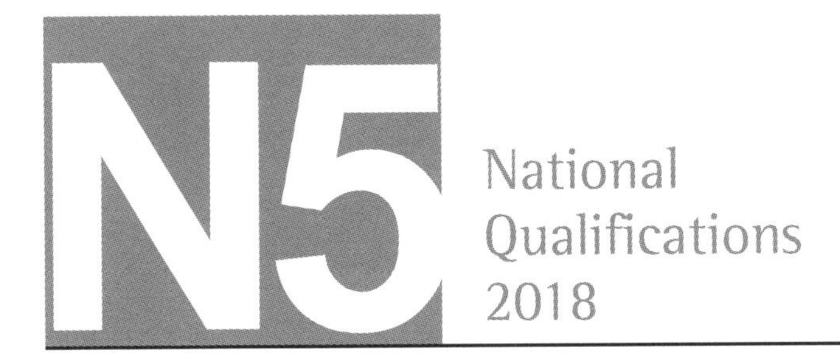

X833/75/21

Geography
Ordnance Survey Map
Item A

TUESDAY, 1 MAY

1:00 PM – 3:20 PM

The colours used in the printing of these map extracts are indicated in the four little boxes at the top of the map extract. Each box should contain a colour; if any does not, the map is incomplete and should be returned to the Invigilator.

1:50 000 Scale
Landranger Series

Extract No 2302/10

Four colours should appear above; if not then please return to the invigilator.

[BLANK PAGE]

DO NOT WRITE ON THIS PAGE

National Qualifications 2018

X833/75/31

Geography
Ordnance Survey Map
Item B

TUESDAY, 1 MAY

1:00 PM – 3:20 PM

The colours used in the printing of these map extracts are indicated in the four little boxes at the top of the map extract. Each box should contain a colour; if any does not, the map is incomplete and should be returned to the Invigilator.

1:50 000 Scale
Landranger Series

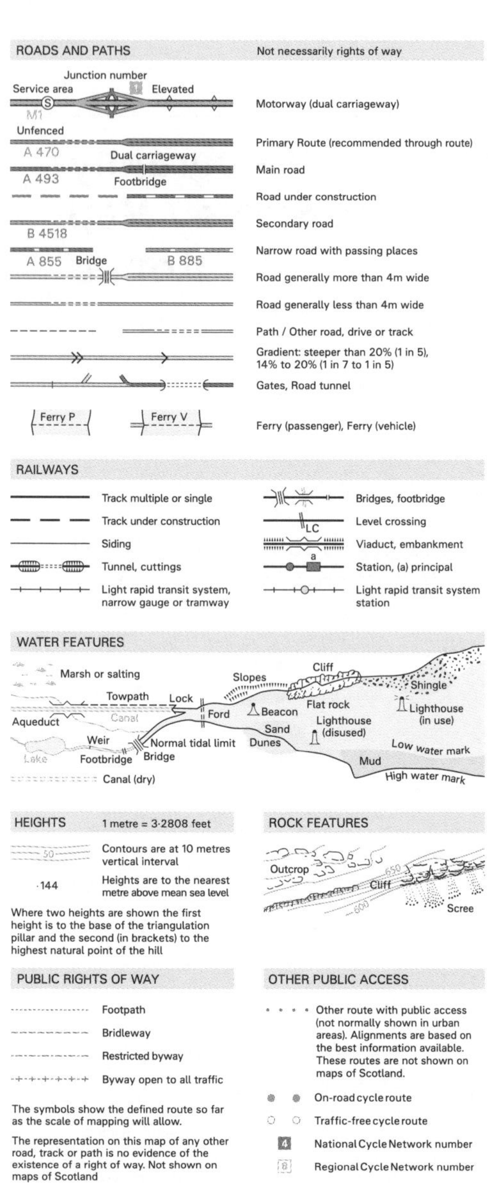

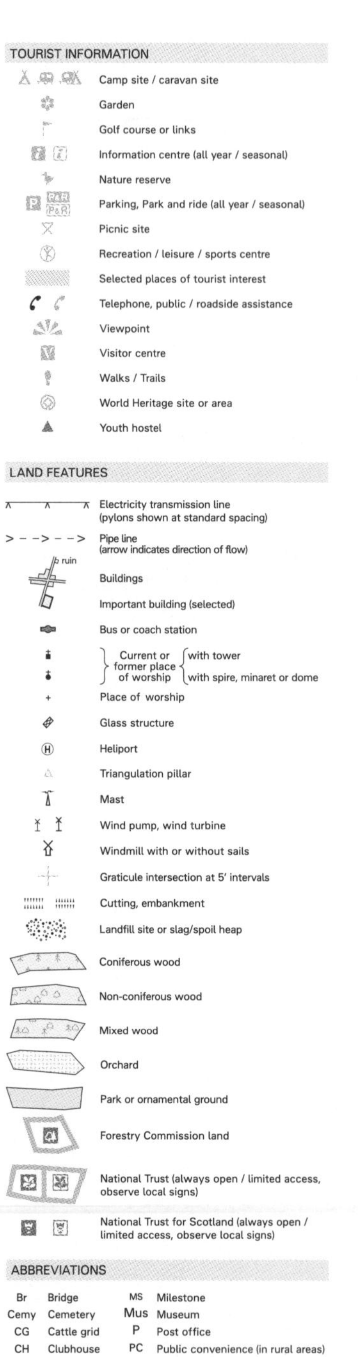

BOUNDARIES

- National
- District
- County, Unitary Authority, Metropolitan District or London Borough
- National Park

ANTIQUITIES

- Site of antiquity
- Battlefield (with date)
- Visible earthwork
- VILLA Roman
- Castle Non-Roman

ABBREVIATIONS

Br	Bridge	MS	Milestone
Cemy	Cemetery	Mus	Museum
CG	Cattle grid	P	Post office
CH	Clubhouse	PC	Public convenience (in rural areas)
Fm	Farm	PH	Public house
Ho	House	Sch	School
MP	Milepost	TH	Town Hall, Guildhall or equivalent

True North
Grid North
Magnetic North
Diagrammatic only

Extract No 2303/164
Four colours should appear above; if not then please return to the invigilator.
Beckley
Elsfield
Stanton St John
Marston
Summertown
New Marston
Headington
New Headington
Sandhills
Risinghurst
Forest Hill
Wheatley
Holton
Waterperry
Great Milton
Horspath
Littleworth
Cowley
Temple Cowley
Littlemore
Blackbird Leys
Sandford-on-Thames
Kennington
South Hinksey
North Hinksey Village
OXFORD DISTRICT
Garsington
Cuddesdon
Denton
Southend
The Platt
Chippinghurst Manor
Stanton Great Wood
Holly Wood
Waterperry Wood
Shotover Country Park
Brasenose Wood
Bagley Wood
Science Park
Stadium
Scale 1: 50 000
2 centimetres to 1 kilometre (one grid square)
Kilometres
Miles
1 kilometre = 0·6214 mile
1 mile = 1·6093 kilometres

[BLANK PAGE]

DO NOT WRITE ON THIS PAGE

NATIONAL 5

2019

X833/75/11 **Geography**

TUESDAY, 28 MAY

1:00 PM – 3:20 PM

Total marks — 80

SECTION 1 — PHYSICAL ENVIRONMENTS — 30 marks

Attempt **EITHER** Question 1 **OR** Question 2

THEN attempt Questions 3 to 7.

SECTION 2 — HUMAN ENVIRONMENTS — 30 marks

Attempt ALL questions.

SECTION 3 — GLOBAL ISSUES — 20 marks

Attempt any **TWO** of the following.

Question 14 — Climate change

Question 15 — Natural regions

Question 16 — Environmental hazards

Question 17 — Trade and globalisation

Question 18 — Tourism

Question 19 — Health

You will receive credit for appropriately labelled sketch maps and diagrams.

Write your answers clearly in the answer booklet provided. In the answer booklet you must clearly identify the question number you are attempting.

Use **blue** or **black** ink.

Before leaving the examination room you must give your answer booklet to the Invigilator; if you do not, you may lose all the marks for this paper.

MARKS

SECTION 1 — PHYSICAL ENVIRONMENTS — 30 marks

Attempt EITHER Question 1 OR Question 2

THEN Questions 3 to 7

Question 1 — Glaciated landscapes/coastal landscapes

(a) Study the Ordnance Survey map extract (Item A) of the Killin area.

Give map evidence which shows that this is an area of **upland glaciated scenery**. 4

Diagram Q1 Cliffs and stacks

(b) Look at Diagram Q1.

Explain the formation of a **stack**.

You may use a diagram(s) in your answer. 4

Now attempt Questions 3 to 7

MARKS

Do not attempt Question 2 if you have already answered Question 1

Question 2 — Rivers and their valleys/upland limestone landscapes

(a) Study the Ordnance Survey map extract (Item A) of the Killin area.

Using grid references **describe the physical** features of the River Lochay **and** its valley between 490369 and 570343. 4

Diagram Q2 Stalactites and stalagmites

(b) Look at Diagram Q2.

Explain the formation of a **stalactite**.

You may use a diagram(s) in your answer. 4

Now attempt Questions 3 to 7

[Turn over

MARKS

Question 3

Diagram Q3 Holiday homes

Study Diagram Q3 and the Ordnance Survey map extract (Item A) of the Killin Area.

A developer has applied for planning permission to build 10 holiday homes in grid square 5833.

Explain the advantages **and** disadvantages of the grid square for this development.

You must use map evidence in your answer. **4**

MARKS

Question 4

Study the Ordnance Survey map extract (Item A) of the Killin Area.

Using map evidence, **describe, in detail,** the attractions of the area shown on the map, for visitors. You may refer to both physical and human attractions in your answer. 5

Question 5

Diagram Q5 Weather report

northerly wind
30 knots
snow showers
sky obscured

Study Diagram Q5.

Draw a weather station circle to show the weather conditions described above. 4

[Turn over

MARKS

Question 6

Diagram Q6A Synoptic chart for 6 am, Sunday 26 February, 2017

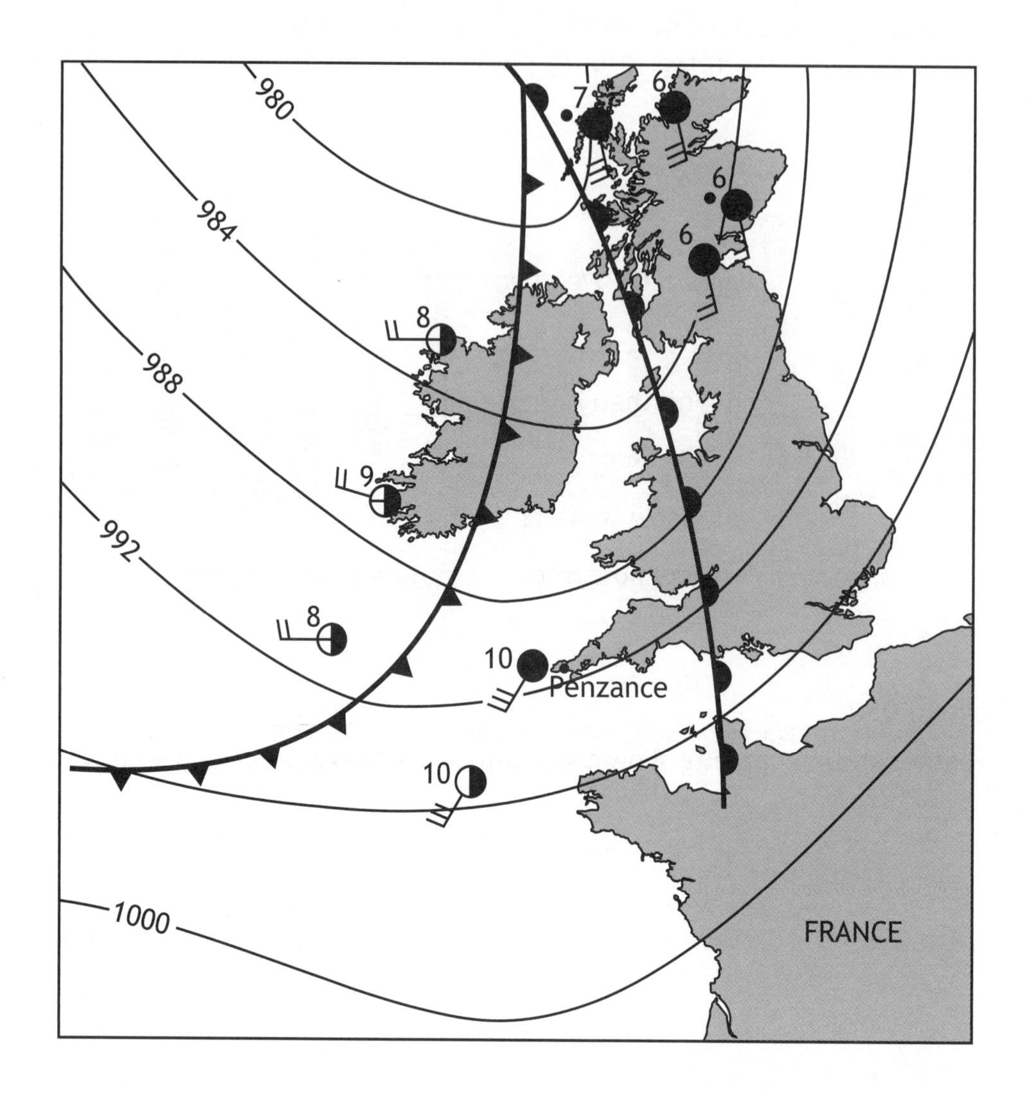

Diagram Q6B Weather forecast for Penzance, Sunday 26 February, 2017

'Sunny intervals will give way to heavy showers. Winds, at first strong, will become lighter and more westerly. Temperatures will drop.'

Study Diagrams Q6A and Q6B.

Using the synoptic chart, **give reasons** for the changes which the weather forecast predicts for Penzance during Sunday 26 February. 4

MARKS

Question 7

Diagram Q7 Wind turbines

Look at Diagram Q7.

Developments such as wind farms can cause land use conflicts.

For a named area you have studied, **explain** why land use conflicts occur there. **5**

[Turn over

MARKS

SECTION 2 — HUMAN ENVIRONMENTS — 30 marks

Attempt ALL questions

Question 8

Diagram Q8 Sketches of urban land use zones

A

B

C

D

Study the Ordnance Survey map extract (Item B) of the Wolverhampton area, and Diagram Q8 (above).

Match each of the urban land use zones shown in Diagram Q8 with the correct grid references below.

Choose from **9004, 9198, 8796, 9101, 8699.** 4

MARKS

Question 9

Study the Ordnance Survey map extract (Item B) of the Wolverhampton area.

Find the old industrial area in grid square 9499. This area is being redeveloped.

Using map evidence, **explain** why this area is attractive for industry. 6

Question 10

Study the Ordnance Survey map extract (Item B) of the Wolverhampton area.

Using map evidence, **describe** methods which have been used to reduce traffic congestion in and around the centre of Wolverhampton.

You should make use of grid references in your answer. 4

[Turn over

MARKS

Question 11

Diagram Q11A Slum housing

Diagram Q11B Quote from development charity website

'By 2030, it is estimated that 1 in 4 people on the planet will live in a shanty town or other informal settlement. Due to population growth and rural-urban migration, these slums are here to stay.'

Look at Diagrams Q11A and Q11B.

Referring to a **developing** world city you have studied, **describe** different ways shanty towns are being improved. 4

MARKS

Question 12

Diagram Q12 Recent developments in UK farming

Look at Diagram Q12

Explain the advantages **and** disadvantages of recent developments in farming in developed countries such as the UK. 6

[Turn over

MARKS

Question 13

Diagram Q13 Population pyramids for Bolivia and the Netherlands

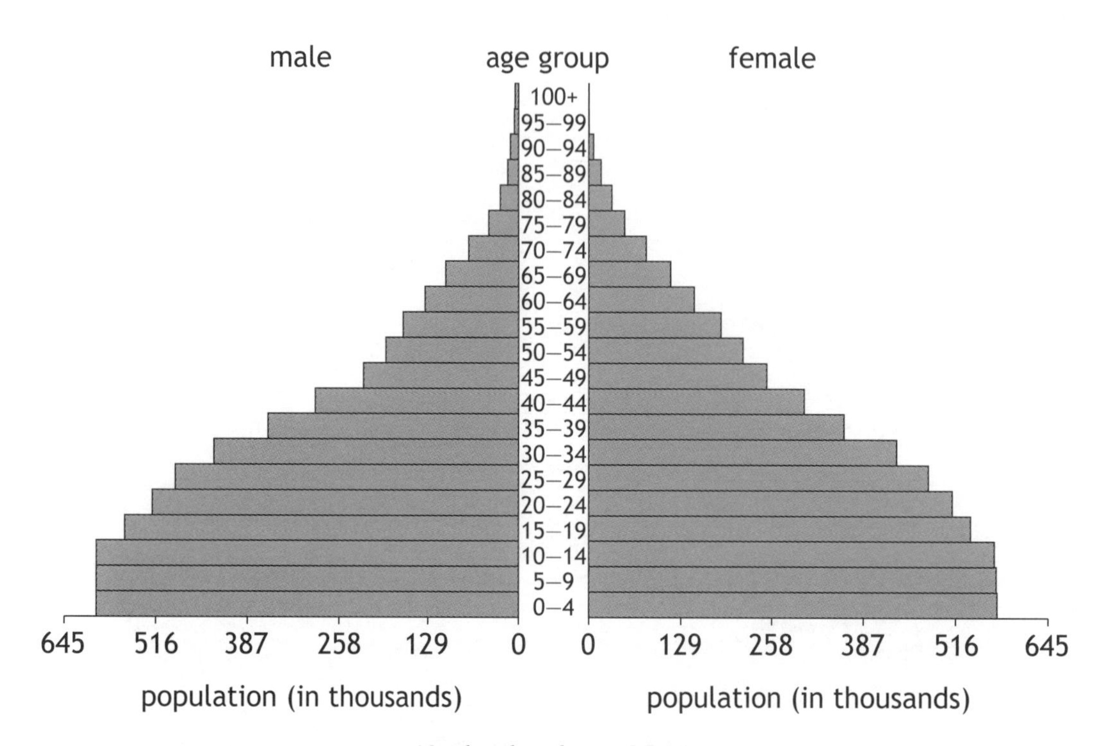

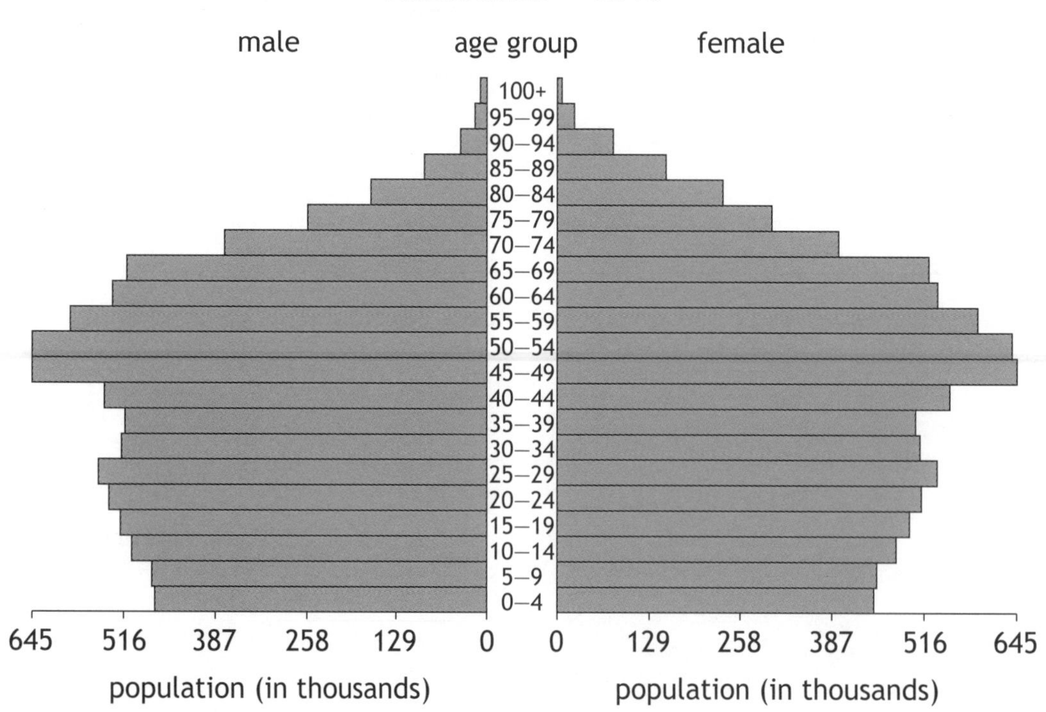

Study Diagram Q13

Explain the differences in the population structures of Bolivia and the Netherlands. You should refer to both birth rates **and** death rates in your answer. 6

SECTION 3 — GLOBAL ISSUES — 20 marks

Attempt any TWO questions

Question 14 — Climate change **(Page 14)**
Question 15 — Natural regions **(Page 16)**
Question 16 — Environmental hazards **(Page 18)**
Question 17 — Trade and globalisation **(Page 20)**
Question 18 — Tourism **(Page 22)**
Question 19 — Health **(Page 24)**

[Turn over

MARKS

Question 14 — Climate change

Diagram Q14A Evidence of climate change

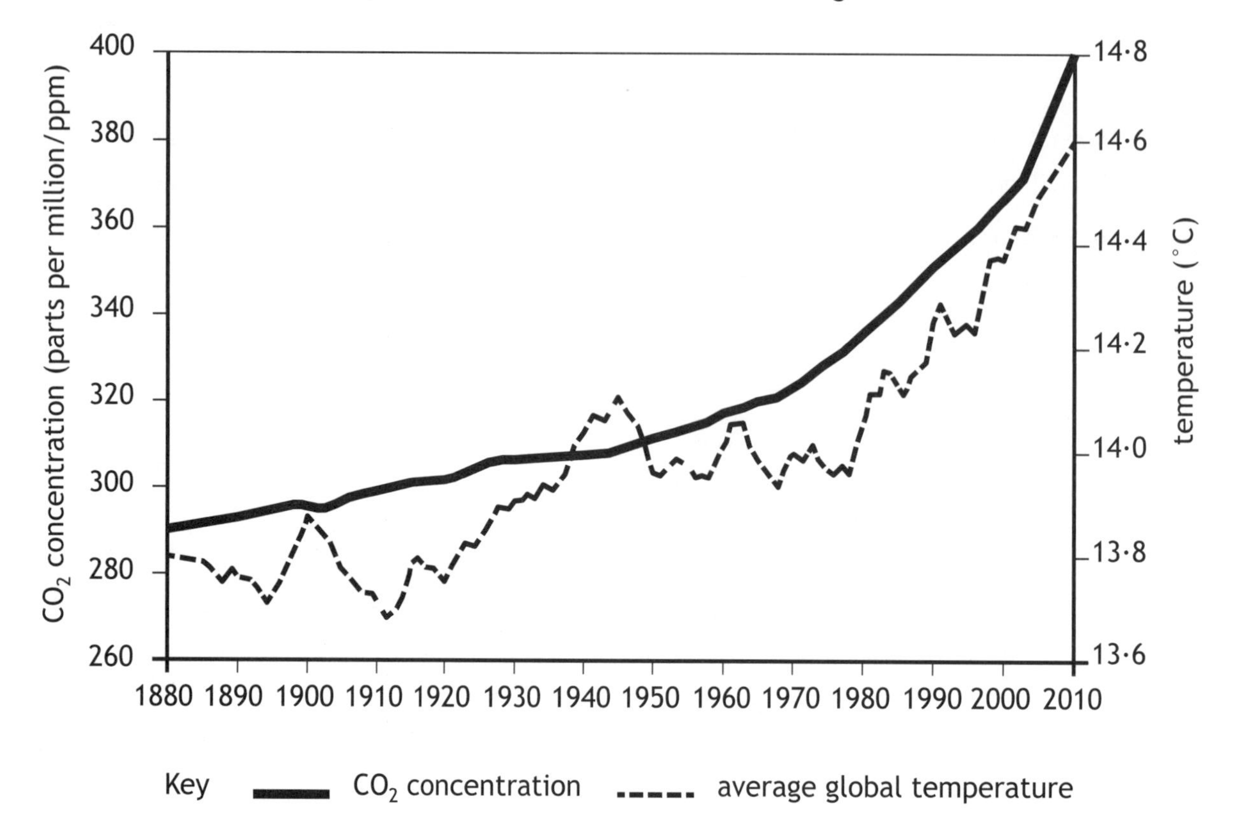

(a) Study Diagram Q14A.

Describe, in detail, the changes in carbon dioxide levels **and** average global temperatures. **4**

MARKS

Question 14 — Climate change (continued)

Diagram Q14B Climate summit 2014 poster

(b) Look at Diagram Q14B.

Explain, in detail, strategies to minimise future climate change. 6

[Turn over

MARKS

Question 15 — Natural regions

Diagram Q15A Deforestation and protected land in the Brazilian rainforest 1995—2015

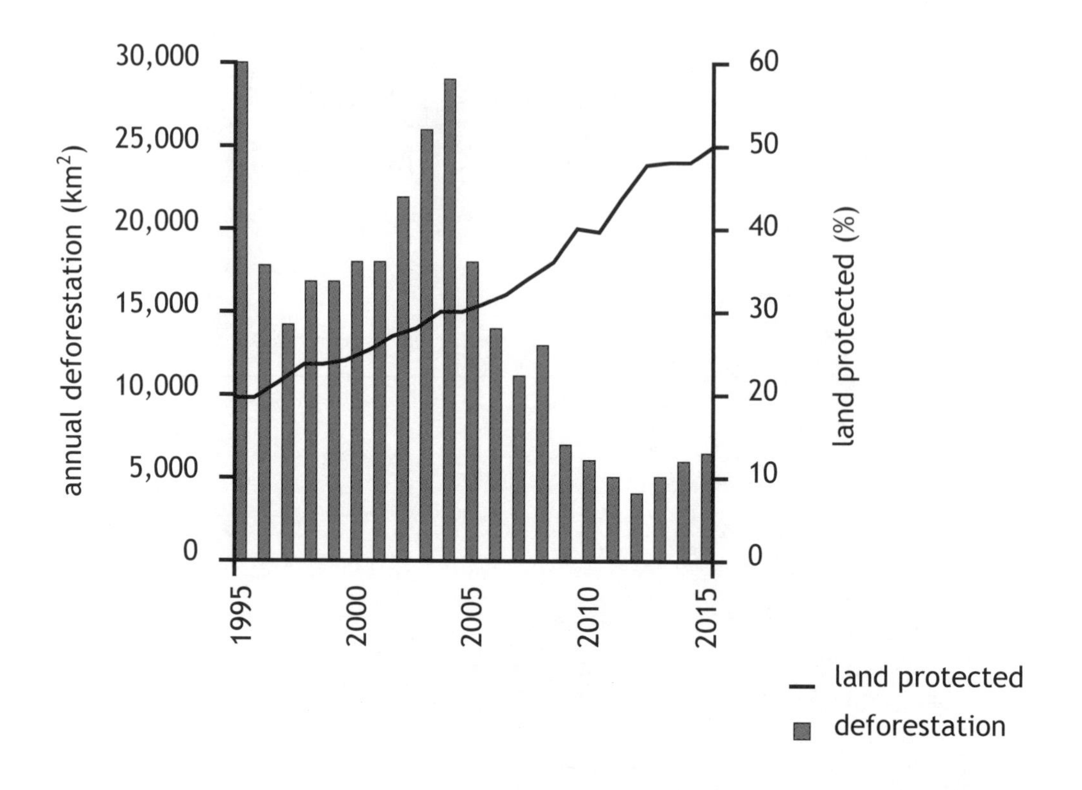

(a) Study Diagram Q15A.

Describe, in detail, changes in deforestation **and** protected land in the Brazilian rainforest. 4

MARKS

Question 15 — Natural regions (continued)

Diagram Q15B An area of tropical deforestation

(b) Look at Diagram Q15B.

For a named area of tropical rainforest you have studied, **describe, in detail,** the strategies used to reduce deforestation. 6

[Turn over

MARKS

Question 16 — Environmental hazards

Diagram Q16A Number of tropical storms in the North Pacific Ocean 1951–2002

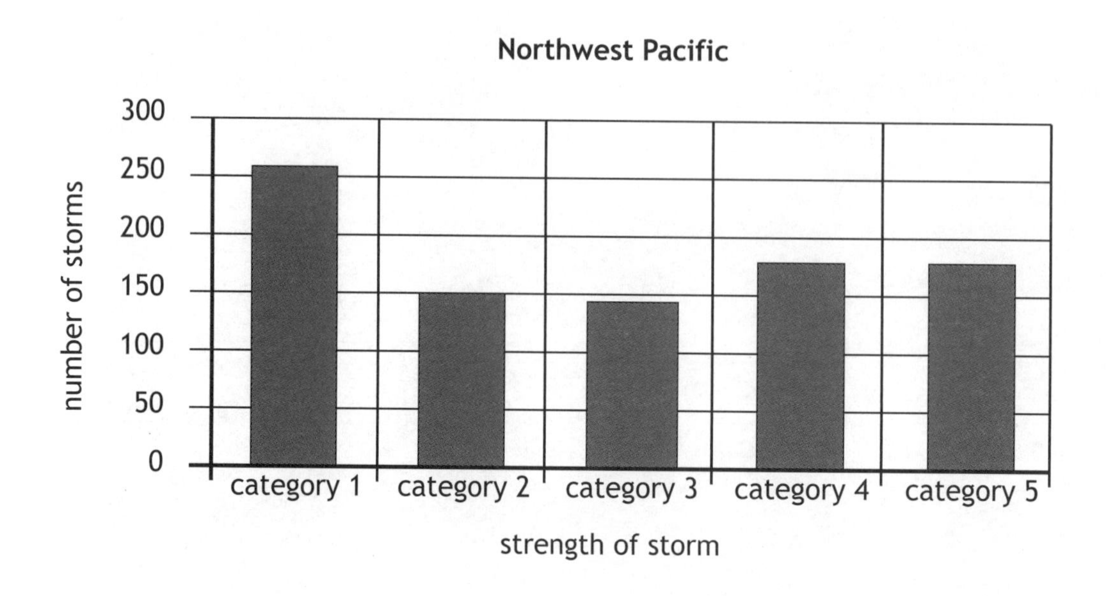

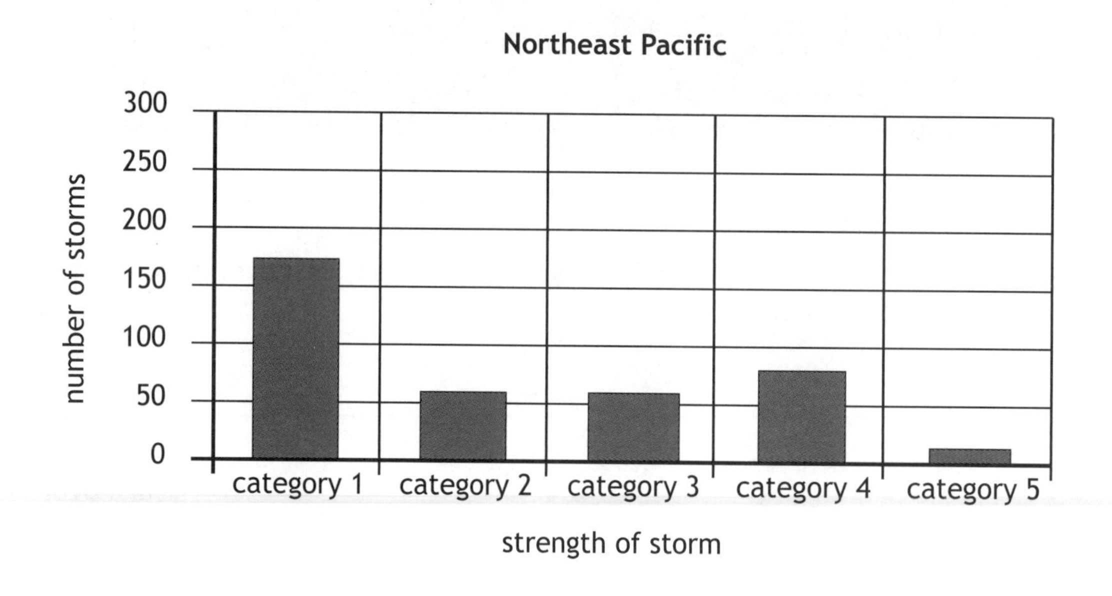

(a) Study Diagram Q16A.

Describe, in detail, the differences in the number of tropical storms in the Northwest and Northeast Pacific Ocean. 4

MARKS

Question 16 — Environmental hazards (continued)

Diagram Q16B A tropical storm

(b) Look at Diagram Q16B.

Explain the formation of a tropical storm.

You may use a diagram(s) in your answer. **6**

[Turn over

MARKS

Question 17 — Trade and globalisation

Diagram Q17A World's top exporting countries 1994–2014

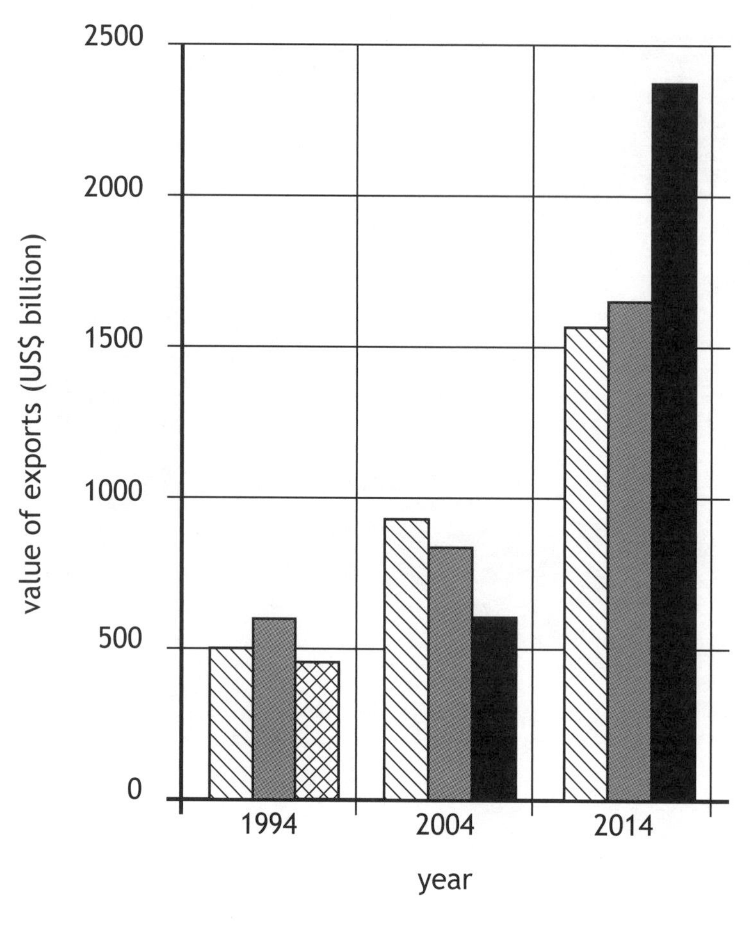

(a) Study Diagram Q17A.

Describe, in detail, the changes in the export value and ranking of the world's top exporting countries. 4

MARKS

Question 17 — Trade and globalisation (continued)

Diagram Q17B Newspaper headline

(b) Look at Diagram Q17B.

Referring to countries you have studied, **describe, in detail**, world trade patterns. 6

[Turn over

MARKS

Question 18 — Tourism

Diagram Q18A International visitors and international visitor spend in Scotland 2005–2017

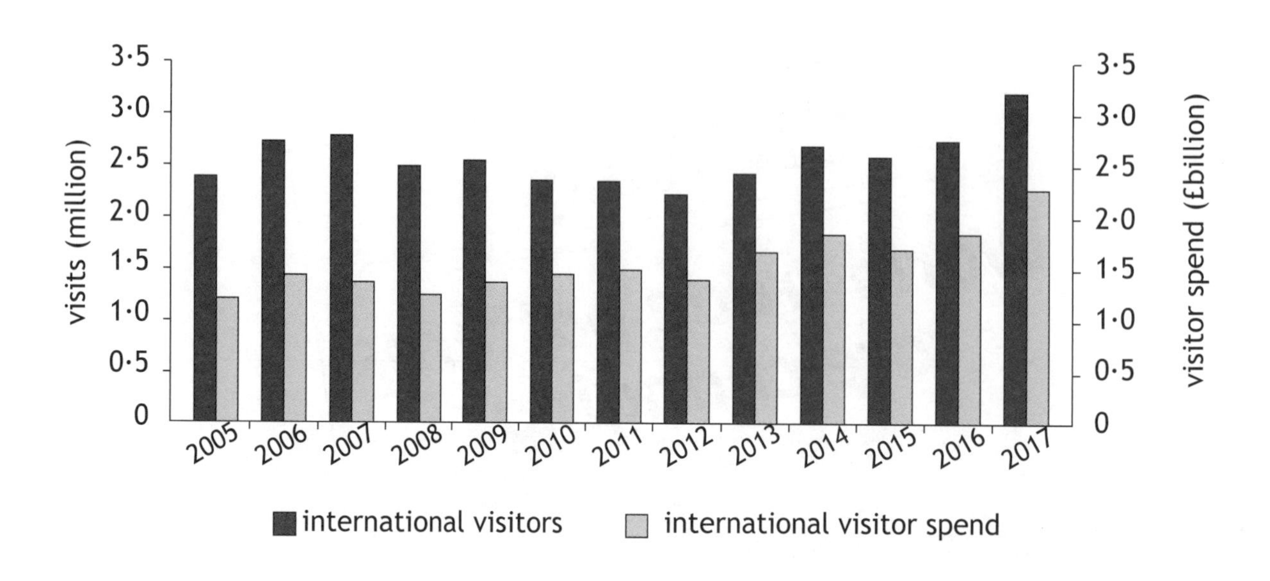

(a) Study Diagram Q18A.

Describe, in detail, the **changes** in the number of international visitors **and** international visitor spend in Scotland between 2005 and 2017. 4

MARKS

Question 18 — Tourism (continued)

Diagram Q18B Newspaper headline

CALEDONIAN HERALD

The International Ecotourism Society encourages responsible travel to natural areas.

(b) Look at Diagram Q18B.

Referring to an area you have studied, **describe, in detail,** the main features of eco-tourism. 6

[Turn over for next question

MARKS

Question 19 — Health

Diagram Q19A Global child mortality rates

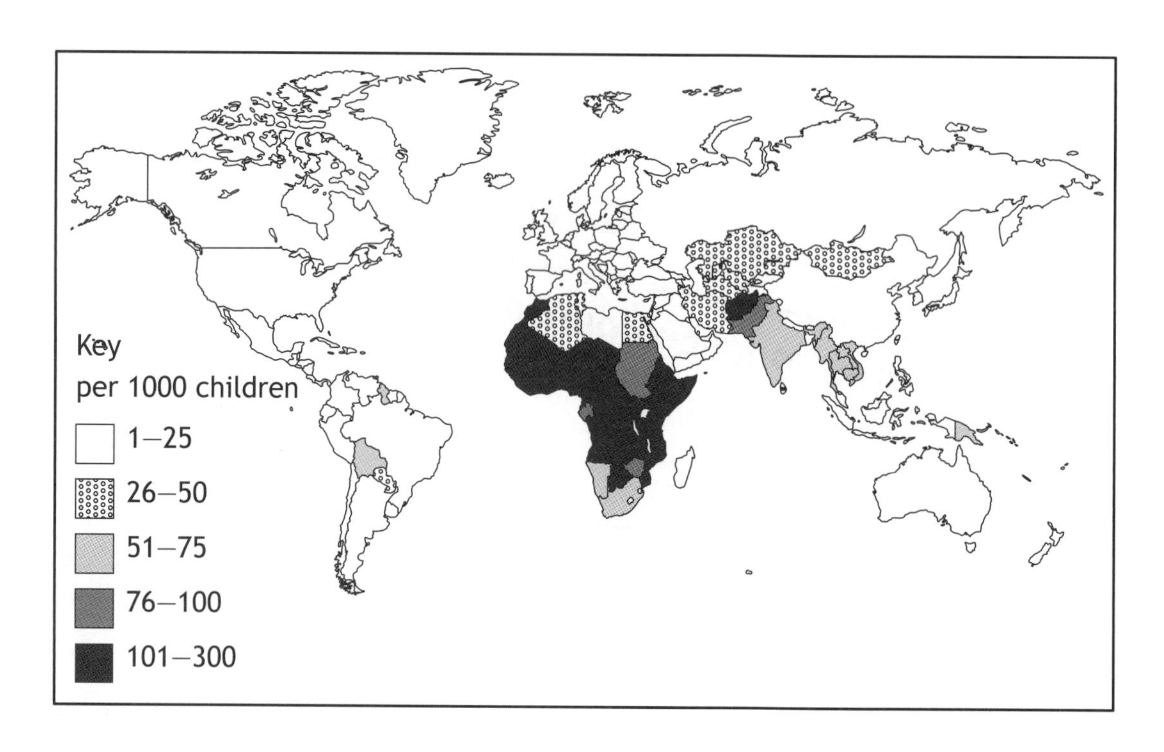

(a) Study Diagram Q19A.

Describe, in detail, child mortality rates across the world. **4**

Diagram Q19B Selected diseases common in the developed world

heart disease

cancer

asthma

(b) Choose **one** disease from Diagram Q19B above.

For the disease you have chosen, **explain** the methods used to manage it. **6**

[END OF QUESTION PAPER]

[BLANK PAGE]

DO NOT WRITE ON THIS PAGE

[BLANK PAGE]

DO NOT WRITE ON THIS PAGE

[BLANK PAGE]

DO NOT WRITE ON THIS PAGE

X833/75/21

Geography
Ordnance Survey Map
Item A

TUESDAY, 28 MAY

1:00 PM – 3:20 PM

The colours used in the printing of these map extracts are indicated in the four little boxes at the top of the map extract. Each box should contain a colour; if any does not, the map is incomplete and should be returned to the Invigilator.

1:50 000 Scale
Landranger Series

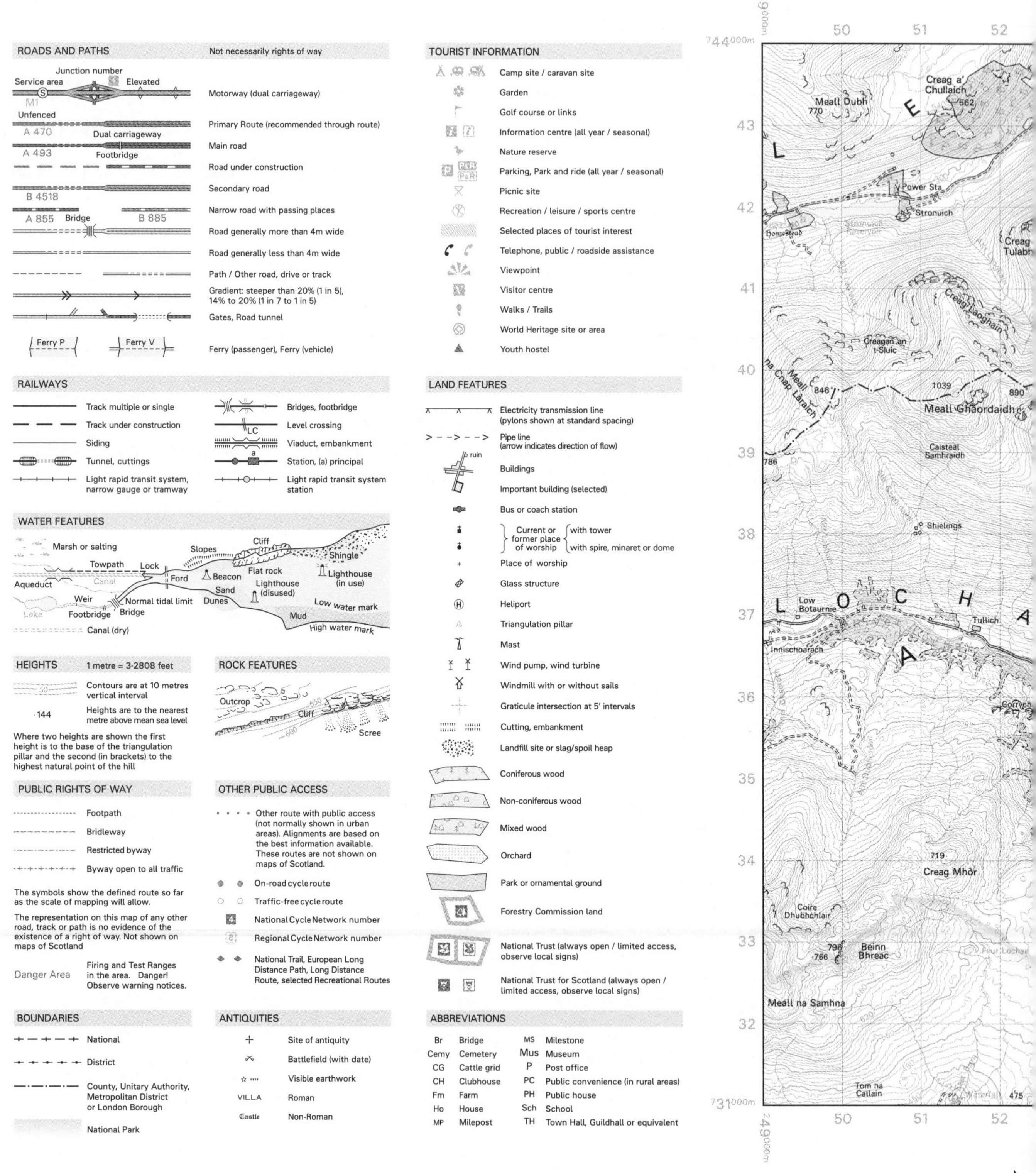

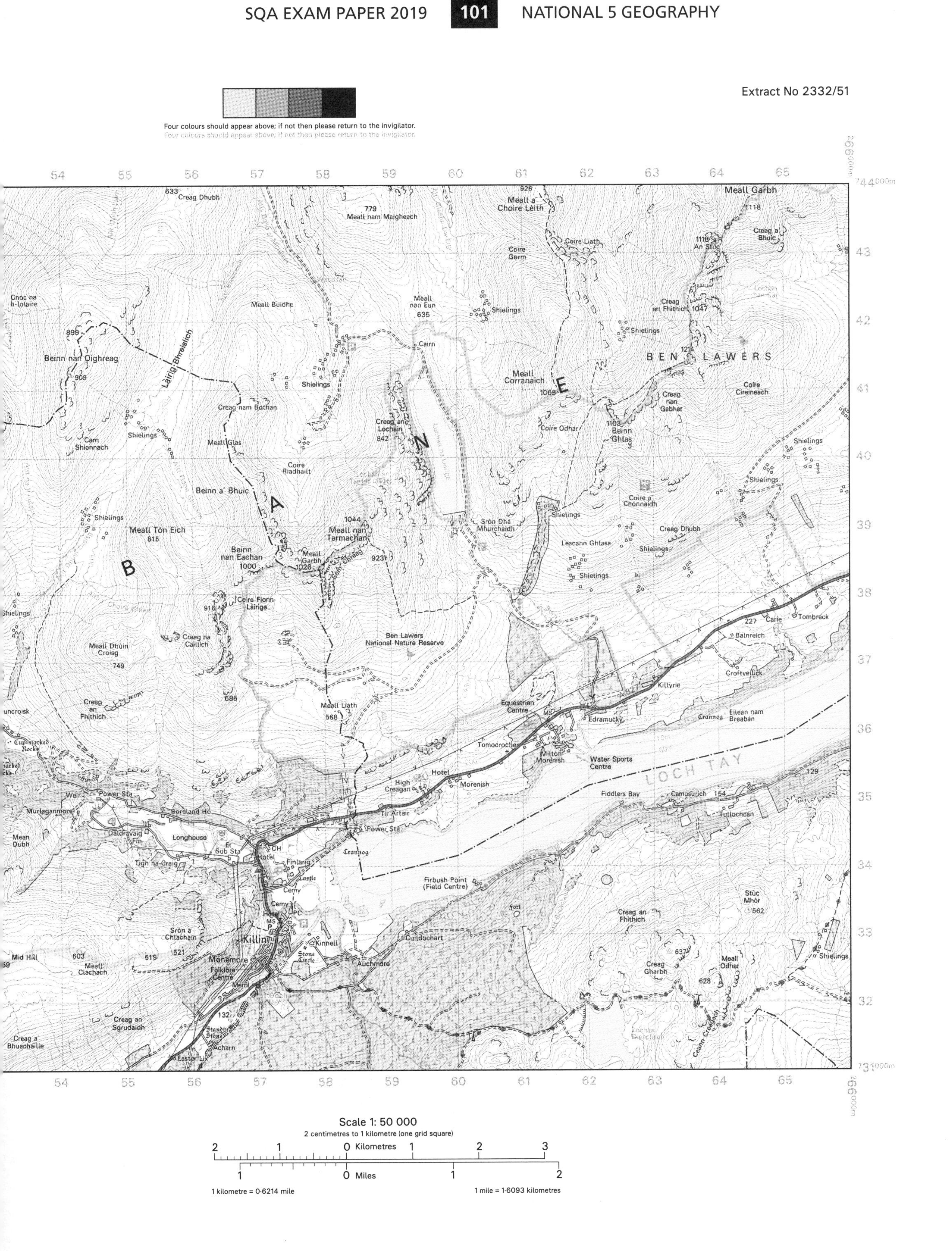
Four colours should appear above; if not then please return to the invigilator.
Extract No 2332/51
Meall Garbh
Meall a' Choire Lèith
Meall nam Maigheach
Creag Dhubh
Coire Liath
Coire Gorm
An Stuc
Creag a' Bhuic
Meall Buidhe
Meall nan Eun
Shielings
Creag an Fhithich
BEN LAWERS
Beinn nan Oighreag
Lairig Bhreislich
Cairn
Meall Corranaich
Coire Cireineach
Creag nan Gabhar
Creag nam Bothan
Creag an Lochain
Coire Odhar
Beinn Ghlas
Meall Glas
Carn Shionnach
Coire Riadhailt
Lochan na Lairige
Beinn a' Bhuic
Coire a' Chonnaidh
Meall Tòn Eich
Meall nan Tarmachan
Sròn Dha Mhurchaidh
Creag Dhubh
Leacann Ghlasa
Beinn nan Eachan
Meall Garbh
B
A
N
E
Coire Fionn-Lairige
Carie
Tombreck
Balnreich
Creag na Caillich
Meall Dhùin Croisg
Ben Lawers National Nature Reserve
Croftvellick
Kiltyrie
Creag an Fhithich
Meall Liath
Equestrian Centre
Edramucky
Eilean nam Breaban
Crannog
Tomocrochar
Milton Morenish
Water Sports Centre
LOCH TAY
Hotel
High Creagan
Morenish
Fiddlers Bay
Camusurich
Tullochcan
Weir
Power Sta
Murlaganmore
Boreland Ho
Daldravaig Fm
Longhouse
El Sub Sta
Tir Artair
Power Sta
Mean Dubh
Tigh na-Craig
Hotel
Finlarig
Castle
Crannog
Firbush Point (Field Centre)
Stùc Mhòr
Cemy
Fort
Sròn a' Chlachain
Killin
Culdochart
Creag an Fhithich
Mid Hill
Meall Clachach
Monemore
Kinnell
Stone Circle
Auchmore
Creag Gharbh
Meall Odhar
Folklore Centre
Meml
Creag an Sgrùdaidh
Creag a' Bhuachaille
Standing Stone
Acharn
Easter Lix
Cuma Creige
Scale 1: 50 000
2 centimetres to 1 kilometre (one grid square)
Kilometres
Miles
1 kilometre = 0·6214 mile
1 mile = 1·6093 kilometres

[BLANK PAGE]

DO NOT WRITE ON THIS PAGE

X833/75/31

Geography
Ordnance Survey Map
Item B

TUESDAY, 28 MAY

1:00 PM – 3:20 PM

The colours used in the printing of these map extracts are indicated in the four little boxes at the top of the map extract. Each box should contain a colour; if any does not, the map is incomplete and should be returned to the Invigilator.

1:50 000 Scale
Landranger Serie

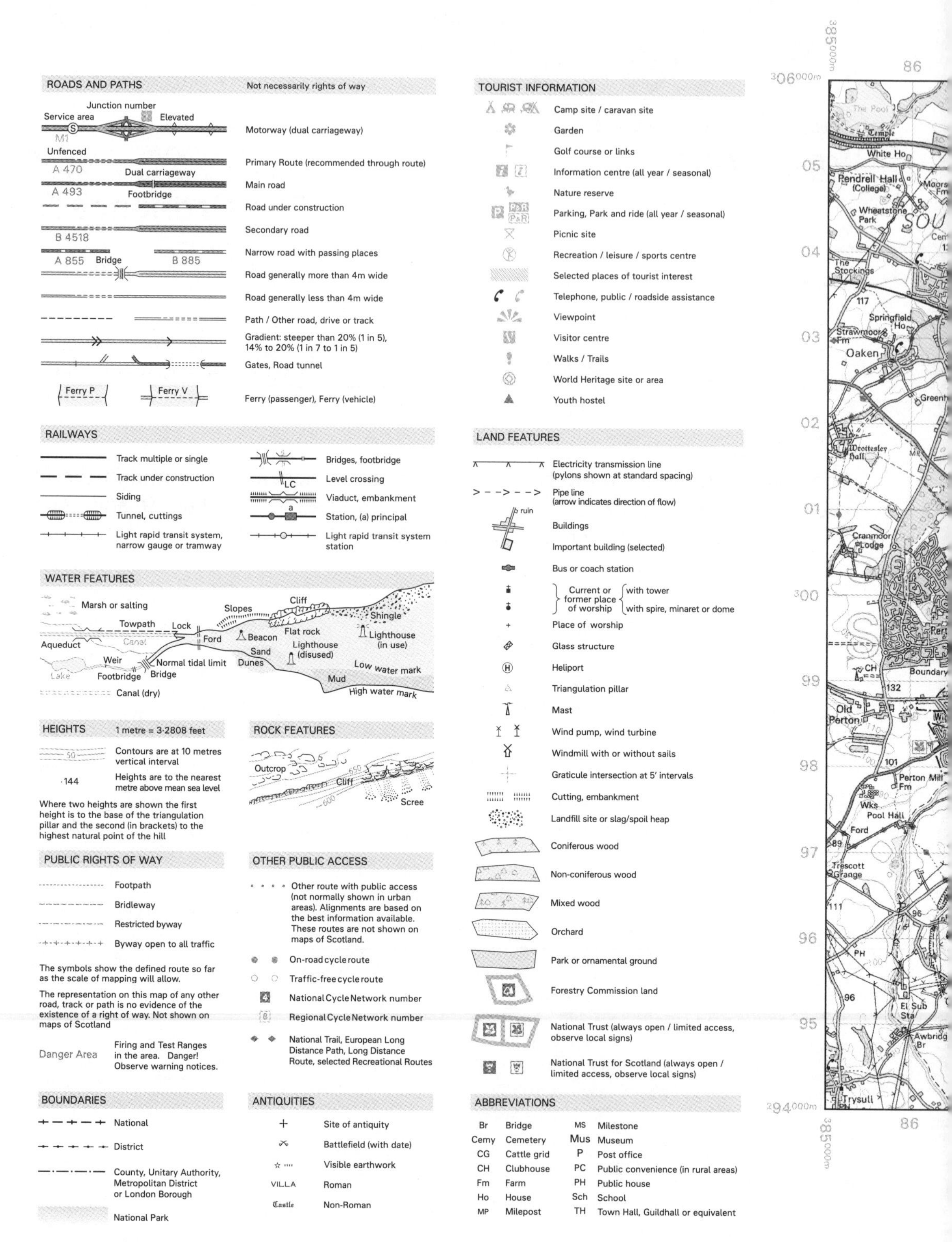

Extract No 2331/139

Four colours should appear above; if not then please return to the invigilator.

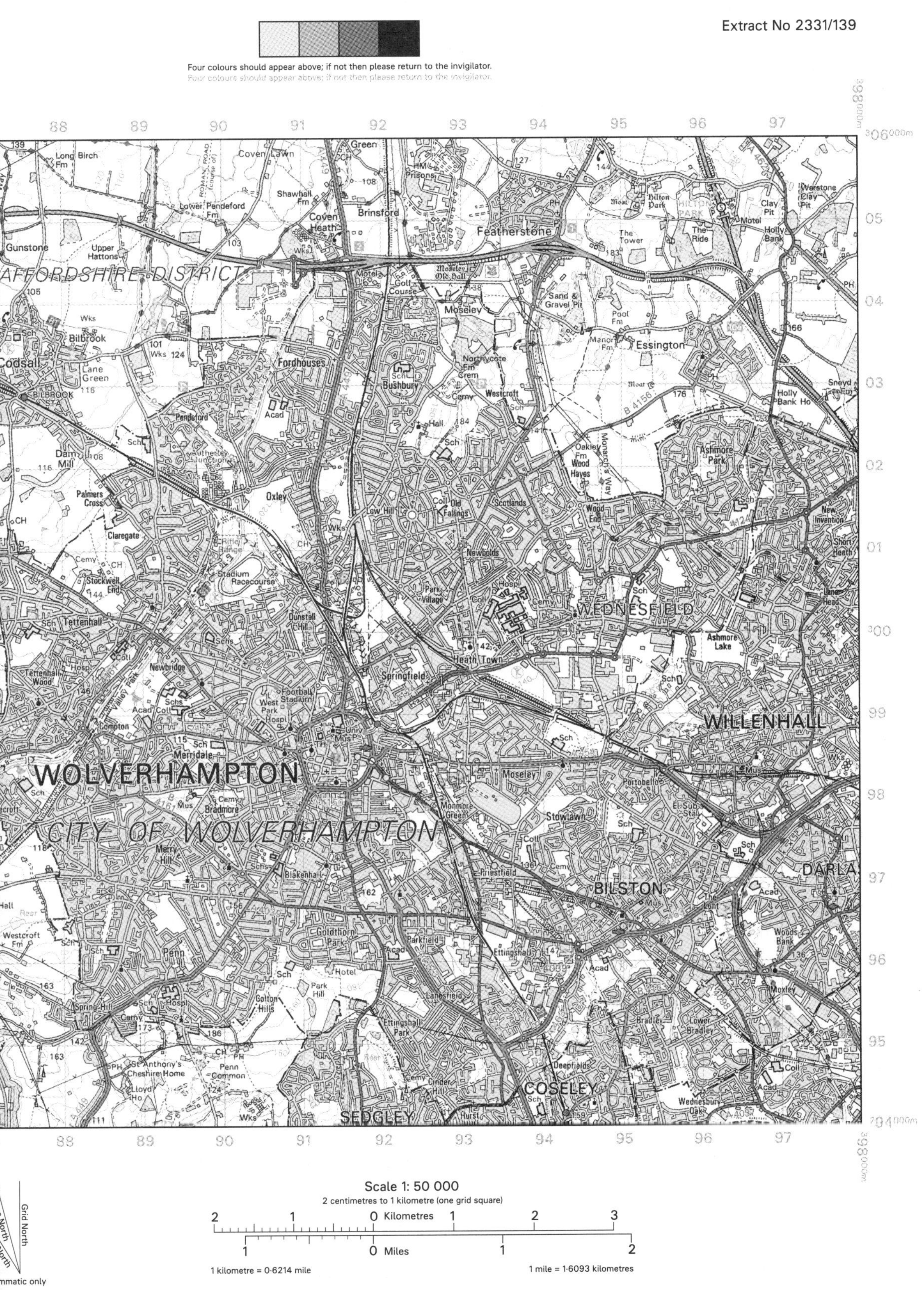

Extract produced by Ordnance Survey Limited 2019.

[BLANK PAGE]

DO NOT WRITE ON THIS PAGE

NATIONAL 5

Answers

ANSWERS FOR

SQA NATIONAL 5 GEOGRAPHY 2019

General Marking Principles for National 5 Geography

Questions that ask candidates to *Describe* . . . (4–6 marks)

Candidates must make a number of relevant, factual points. These should be key points. The points do not need to be in any particular order. Candidates may provide a number of straightforward points or a smaller number of developed points, or a combination of these.

Up to the total mark allocation for this question:

- **One mark** should be given for each accurate relevant point.
- **Further marks** should be given for development and exemplification.

Question: Describe, in detail, the effects of two of the factors shown. (Modern factors affecting farming).

Example:

New technology has led to increased crop yields *(1 mark)*, leading to better profits for some farmers *(second mark for development)*.

Questions that ask candidates to *Explain* . . . (4–6 marks)

Candidates must make a number of points that make the process/situation plain or clear, for example by showing connections between factors or causal relationships between events or processes. These should be key reasons and may include theoretical ideas. There is no need for any prioritising of these reasons. Candidates may provide a number of straightforward reasons or a smaller number of developed reasons, or a combination of these. The use of the command word "explain" will generally be used when candidates are required to demonstrate knowledge and understanding. However, depending on the context of the question the command words "give reasons" may be substituted.

If candidates produce fully labelled diagrams they may be awarded up to full marks if the diagrams are sufficiently accurate and detailed.

Up to the total mark allocation for this question:

- **One mark** should be given for each accurate relevant point.
- **Further marks** should be given for developed explanations.

Question: Explain the formation of a U-shaped valley.

Example:
A glacier moves down a main valley which it erodes (*1 mark*) by plucking, where the ice freezes on to fragments of rock and pulls them away (*second mark for development*).

Questions that ask candidates to *Give reasons* . . . (4–6 marks)

Candidates must make a number of points that make the process/situation plain or clear, for example by showing connections between factors or causal relationships between events or processes. These should be key reasons and may include theoretical ideas. There is no need for any prioritising of these reasons. Candidates may provide a number of straightforward reasons or a smaller number of developed reasons, or a combination of these. The command words "give reasons" will generally be used when candidates are required to use information from sources. However, depending on the context of the question the command word "explain" may be substituted.

Up to the total mark allocation for this question:

- **One mark** should be given for each accurate relevant point.
- **Further marks** should be given for developed reasons.

Question: Give reasons for the differences in the weather conditions between Belfast and Stockholm.

Example:
In Stockholm it is dry, but in Belfast it is wet because Stockholm is in a ridge of high pressure whereas Belfast is in a depression (*1 mark*). Belfast is close to the warm front and therefore experiencing rain (*second mark for development*).

Questions that ask candidates to *Match* (3–4 marks)

Candidates must match two sets of variables by using their map interpretation skills.

Up to the total mark allocation for this question:

- **One mark** should be given for each correct answer.
- **Question:** Match the letters A to C with the correct features.

Example: A = Forestry (*1 mark*)

Questions that ask candidates to *Give map evidence* (3–4 marks)

Candidates must look for evidence on the map and make clear statements to support their answer.

Up to the total mark allocation for this question:

- **One mark** should be given for each correct statement.

Question: Give map evidence to show that part of Coventry's CBD is located in grid square 3379.

Example: Many roads meet in this square (*1 mark*).

Questions that ask candidates to *Give advantages and/or disadvantages* (4–6 marks)

Candidates must select relevant advantages or disadvantages of a proposed development and show their understanding of their significance to the proposal. Answers may give briefly explained points or a smaller number of points which are developed to warrant further marks.

Up to the total mark allocation for this question:

- **One mark** should be given for each accurate relevant point.
- **Further marks** should be given for developed points.
- Marks should be awarded for accurate map evidence.

Question: Give either advantages or disadvantages of this location for a shopping centre. You must use map evidence to support your answer.

Example: There are roads and motorways close by allowing the easy delivery of goods (*1 mark*) and access for customers (*1 mark for development*), eg the A46, M6 and M69.

NATIONAL 5 GEOGRAPHY 2017 SPECIMEN QUESTION PAPER

1. (a) Corrie with lochan in squares 7921 & 8021 (1). Scree slopes on south side of corrie (1). U-shaped valley at 8414 (1). Pyramidal peak (Pen-y-Fan) at 012215 (1). Arete at 016213 (1). Corrie with lochan (Llyn Cwn Llwch) at 0022 (1).

   Or any other valid point.

   (b) Glacier forms in corrie/north facing slope and moves downhill due to gravity (1). Eroding sides and bottom of valley (1) through plucking and abrasion (1). Action makes valley sides steeper and valley deeper (1). When glacier retreats a deep, steep, flat-floored U-shaped valley left behind (1). Original river in valley now seems too small for wider valley and is known as misfit stream (1).

   Or any other valid point.

2. (a) From point 905175 river is flowing south (1) down steep-sided V-shaped valley (squares 9017 & 9016) (1). River is joined by tributaries from west such as at 906166 (1). Confluence at 911153 (1) and from this point river gets wider (1). Meander at 911152 (1). At least 3 waterfalls marked in square 9009 (1). In this square river is flowing south-west (1).

   Or any other valid point.

   (b) At river meander, water pushed towards outside of bend causing erosion (1), by processes such as corrosion or hydraulic action (1). Slower flow of water on inside bend causes deposition (1). Over time erosion narrows neck of meander (1). In time, usually during a flood, river will cut right through neck (1). Fastest current is now in centre of river and deposition occurs next to banks (1) eventually blocking off meander to leave ox-bow lake (1).

   Or any other valid point.

3. Answers will vary depending upon the land uses chosen.

   Tourism and recreation examples: Tourists able to visit show caves at Dan-yr-Ogof in 8316 (1) other attractions nearby which might be of interest such as Shire Horse Centre and public house in 8416 (1). Camp site in this square where would be able to stay (1). Lots of opportunities for outdoor enthusiasts such as walking Brecon Beacons Way (7922) (1) nature reserves such as in 8615 where visitors may see rare wildlife species (1). Climbers could tackle the cliffs in Pen-y-Fan (0121).

   Farming: Lots of mountainous land suitable for hill sheep farming (1) sheep able to survive in the colder, windy and wet conditions (1). Farms such as Coed Cae Ddu in 9510 have good road connections, only about 2 kilometres from an A class road, giving good access to markets (1) patches of woodland which provide shelter for the ewes, especially at lambing time (1). Other farms such as Pwllcoediog (8416) able to benefit from high numbers of visitors by earning extra income from bed and breakfast (1).

   Industry: Limestone areas such as Brecon Beacons are sometimes used for extraction of limestone (1). Quarries could be built here as there is limestone and an A class road (A4067) nearby for material to be transported (1). Opencast working shows evidence of industry, grid reference 8211 (1). Works located at 847107 where land is flat so easy to build on (1). Good communication with two main roads A4109 and A4221 close by, also rail link into works (1). Small settlements close by like Dyffryn Cellwen where workers could be found (1).

4. (a) Cape Wrath has north wind whereas Banbury has west wind (1). 35 knots at Cape Wrath but calmer in Banbury at 15 knots (1). Dry in Banbury but snow showers at Cape Wrath (1). 6 oktas cloud cover at Cape Wrath but only 2 oktas over Banbury (1). Temperature at Cape Wrath is much colder at 2°C, while at Banbury is 11°C (1).

   (b) Should not go walking in hills as cold front about to arrive in area (1) which will bring heavy rain showers (1). Will also cause temperature to drop close to freezing point and could be snow (2). If not properly equipped could suffer from the cold and get hypothermia (1) especially as isobars are close together resulting in high wind chill (1). If heavy snow or low cloud they could lose their way easily and need to be rescued (1) these conditions are life-threatening and they should wait for a better day (1).

   Or any other valid point.

5. Tropical continental air mass will bring hot, dry weather in summer which could result in droughts (1). Might need to be hosepipe bans (1). Grass might wither and die causing problems for livestock farmers (1). Ice cream sales might rise (1) as people make most of sunny weather and head for beach (1). Could be very hot and difficult to do physical work outside (1). Heavy rain from thunderstorms might cause flash floods (1).

   Or any other valid point.

6. Answers will vary depending upon the land uses chosen.

   Examples of problems between tourists and farmers: In Cairngorms, tourists can disrupt farming activities as walkers leave gates open, allowing animals to escape (1). Tourists' dogs can worry sheep if let off the lead (1). Stone walls are damaged by people climbing over instead of using gates/stiles (1). Solved by putting in kissing gates or stiles so that people don't have to open gates (1). Noisy tourists can disturb sheep especially during breeding season (1). Farmers may restrict walkers' access at certain times eg lambing season (1). Farm vehicles can slow up tourist traffic on roads (1) parked cars on narrow country roads can restrict movement of large farm vehicles (1). Some of these problems can be resolved by educating public through methods such as publicising Country Code (1).

   Examples of problems between industry and tourists: Tourists want to see beautiful and unusual scenery of Yorkshire Dales but quarries spoil natural beauty of landscape (1). Lorries used to remove stone endanger wildlife and put visitors off returning to area (1). This threatens local tourist-related jobs eg in local restaurants (1). Large lorries needed to remove quarried stone cause air pollution and dust which spoils atmosphere for tourists (1). Quarry companies have covered vehicles with tarpaulins to try and reduce amount of dust (1). Lorries cause traffic congestion on narrow country roads which slows traffic and delays drivers (1). Some quarries have reduced number of lorries by sending limestone by rail instead (1). Peace and quiet for visitors is disturbed by blasting of rock (1). Quarry companies limit frequency and times of blasting to try to reduce impact on local communities (1). Some wildlife habitats may also be disturbed by removal of rock (1).

   Or any other valid point.

7. (a) Main roads lead into this square (1) there is a bus station (1) and two railway stations (1) tourist information centre (1) several churches (1) museum (1).

Or any other valid point.

(b) Land is flat so easy to build on (1) space available for expansion (1) eg expansion of motor works at 163823 (1). Good transports links like M42 allowing people and products access to and from area (1). Rail link with Birmingham International rail station gives easy access to airport (1). Many road junctions and intersections connecting area to other areas and less traffic congestion as is away from Birmingham city centre (2). Land is on edge of Birmingham so will be cheaper encouraging housing estates like Sheldon to be built (1). Cheaper land allows houses to be bigger with cul-de-sacs, gardens etc. (1) houses can provide a source of labour for the airport, motor works and business park (1).

Or any other valid point.

(c) **Area Y:** Some quieter roads such as cul-de-sacs which would be peaceful area to live in (1). Balsall Heath very close to cricket ground (0684) which would be good for recreation (1) and Cannon Hill Country Park (0683) also close by which would give family easy access to pleasant place to walk (1). Children might also enjoy seeing wildlife at Nature Centre in park (1). Area Y is close to centre of Birmingham which might make it easier for parents to get to work if their jobs are there (1) will also be convenient for shopping as lots of variety in CBD (1). Appears to be industrial areas in Area Y (8410) which might provide job opportunities for parents, conveniently close to where they will live (1).

Area Z: Further out of city, land prices will be cheaper and may be able to afford better house (1). Evidence of lots of modern housing estates with curvilinear road patterns which will be nice environment to live in as will be more garden space (1) and less traffic, making it safer for families (1). Two schools in Area Z which mean children will not have long journey to get there (1). Two golf courses in Area Z, providing recreation opportunities for Russell family (1) and able to get outdoor exercise easily by walking along North Worcestershire Path which passes through Area Z (1). Good transport links into Birmingham for shopping or jobs via main A435 road which leads into CBD and also nearby Park & Ride (1077) next to station where they could travel by train into centre (1). Lots of open space and areas of woodland make the environment/air quality better than many other parts of city (1).

Or any other valid point.

8. Pesticides reduce disease producing better crops (1) and surplus to trade (1). Fertilisers increase crop yields (1) leading to better profits for some farmers (1) which can lead to increase in standard of living (1). Mechanisation means less strenuous work for farmer (1) and is quicker and more efficient (1). GM crops produce greater yield and are disease resistant so make a greater profit for farmer (1) can reduce cost to farmer of applying pesticides (1) and reduce risk to his health (1). Growing demand for biofuels means higher crop prices and can result in farmer getting higher income (1) and create employment (1).

Or any other valid point.

9. Stage 2 birth rate was still very high, whereas death rate fell quite quickly – caused rapid rise in total population as people were living longer (1). Death rates fell because clean water supplies introduced, reducing spread of disease (1) and at same time proper sewage systems being built which meant water supplies no longer contaminated, reducing number of people falling ill and dying (2). Advances in medicine such as introduction of penicillin helped keep death rates low, as people could be treated for and cured from illnesses which might have killed them in past (2). Birth rates were still high as people were used to idea many children may not survive until adulthood (1) and also children were required to go out and work because of poverty (1). Stage 3 birth rates started to drop much faster as people realised infant mortality was falling and no longer needed to have extra children as insurance policy (2). Standard of living had improved and wasn't necessary to have lots of children to earn income for family any more (1). Also education about family planning was more common (1) and availability and variety of different methods of contraception was better (1). Falling birth rate meant that rate of population increase started to slow down (1). Stage 4 birth rate fallen so low that in some countries is below death rate and so overall population is falling (1). Japan is example of this (1).

Or any other valid point.

10. **Life expectancy:** Very useful development indicator as shows that people in developed countries such as Finland live much longer than in developing countries such as Chad (1). Likely to be because standard of living in Finland is much better (1) and will be much better hospitals, more doctors and more money to spend on medicine (1) as Finland is wealthy developed country which can afford to pay for all of this (1). In Chad, people will have very hard physically demanding lives which may lead to shorter life expectancy (1). They may also live shorter lives on average because of poor nutrition, food shortages and famine (1).

Percentage of workforce employed in agriculture: Can tell you a lot about a country because if a very high proportion of workforce in agriculture shows less developed country (1). Whereas in developing countries such as Mali, 80% of workers are employed in farming because mostly subsistence farmers who have to grow own food (1). Few other places that can get food from or simply can't afford to buy it (2). Also few other industries for people to get jobs in as country is less developed and there is lack of money to invest in setting up new businesses (2). Developed countries such as the Netherlands have very efficient farming industries which require very few workers (1); their economy is highly developed meaning most people are employed in many other jobs and industries which are available and which provide higher incomes than farming (2).

Or any other valid point.

11. (a) Overall trend is that amount of Arctic Sea ice has decreased between 1979 and 2013 (1) from (around) 7 million square km to (about) 5 million square km (1). Fluctuation in extent of sea ice in certain years (1) eg amount of sea ice increased from 3.75 million square km in 2012 to 5 million square km in 2013 (1). Between 2006 and 2007 was a sharp decrease (1) from 6 million square km to 4.25 million square km (1).

Or any other valid point.

(b) Increased temperatures causing ice caps to melt so Polar habitats beginning to disappear (1). Melting ice causes sea levels to rise (1) threatening coastal settlements (1). Increase in sea temperatures causes water to expand, compounding problem of flooding (1). Global warming could also affect weather patterns, leading to more droughts (1) crop failures and problems with food supply (1). Flooding, causing the extinction of species (1) and more extreme weather, eg tropical storms (1). Tourism problems will increase as will be less snow in some mountain resorts (1). Global warming could threaten development of developing countries as restrictions on fossil fuel use may be imposed to slow rate of increasing CO_2 levels (1). In UK, tropical diseases like malaria may spread as temperatures rise (1). Plant growth will be affected and some species will thrive in previously unsuitable areas (1). Higher temperatures may cause water shortages (1).

Or any other valid point.

12. (a) High rates of deforestation occur in Brazil, DR Congo and Indonesia (1). High rates are also prevalent in areas such as Mexico and most of South America (1). High levels of loss more common in developing countries (1). Moderate levels common throughout Europe, northern Africa and Canada (1). Low rates common throughout USA, China, India and Australia (1).

Or any other valid point.

(b) **Management strategies include:** Habitat Conservation Programmes sometimes established in tundra environments to protect unique home for tundra wildlife (1). In Canada and Russia, many tundra areas protected through national Biodiversity Action Plan (BAP) (1). BAP is internationally recognised programme designed to protect and restore threatened species and habitats (1). Reducing global warming is crucial to protecting tundra environment because heating up of Arctic areas is threatening existence of environment (1). Most governments have promised to reduce greenhouse gases by signing up to Kyoto Protocol (1). Many countries have invested heavily in alternative sources of energy such as wind, wave and solar power. These sources of energy are renewable and more environmentally friendly than burning fossil fuels, which increase carbon emissions and global warming (2). Some oil companies now schedule construction projects for the winter season to reduce environmental impact (1). Projects work from ice roads, which are built after ground is frozen and snow covered. This limits damage to sensitive tundra (1). Some oil companies locate polar bear dens using infrared scanners and do not work within 1.6 kilometres of these dens (1). Number of Arctic research programmes, such as International Association of Oil & Gas Producers' joint industry programme on Arctic oil spill response technology (1). This programme attempts to increase effectiveness of dispersants in Arctic waters, oil spill modeling in ice and use of remote sensors above and under water (2). Many companies operate sophisticated systems to detect leaks (1). Many companies work with local communities to understand and manage potential local impacts of their work (1). Many countries have set up national parks such as Arctic National Wildlife refuge in Alaska to protect endangered animals in tundra (1). Trans-Alaskan pipeline is raised up on stilts to allow Caribou to migrate underneath (1).

Or any other valid point.

13. (a) Over last 100 years, number of eruptions has increased from 43 in 1910s to 70 in 2010s (1). Apart from decades of 1920s, 1970s and 1990s, amount of volcanic activity in each decade increased (1). Least number of eruptions was in 1920s with only 31 (1). Big drop between 1910s and 1920s with a drop of 12 eruptions (1). Also in 1990s there were 12 fewer eruptions than 1980s (1). Biggest increase between 1990s and 2000s with 13 more eruptions (1). Decades with greatest number of eruptions were 1980s, 2000s and 2010s at 66, 67 and 70 (1).

Or any other valid point.

(b) **For Pico de Fogo volcano:** Heat from lava flows set fire to main settlements destroying two villages as well as forest reserve (2) endangering the vegetation and animal habitat (1). Around 1,500 people forced to abandon homes before lava flow reached villages of Portela and Bangeira on Fogo island (1). More than 1,000 people evacuated from Cha das Caldeiras region at foot of volcano to ensure safety and prevent injuries (1). Airport was closed, as ash filled sky, to prevent risk of planes crashing (1). Buildings and records were destroyed resulting in some of history of area being lost (1). Roads and transport routes destroyed, affecting tourist industry on island (1). Volcano destroyed agricultural land which resulted in loss of fertile land (1) decreasing ability of area to produce crops (1) and support local population (1). Tourism might increase as volcano becomes tourist attraction improving economy of the island (1).

Or any other valid point.

14. (a) Europe dominated world trade exports with around 43% in 2005 (1). Dropped to around 38% in 2010 (1). Europe still largest exporter in 2010 (1). Asia had second largest regional share of world trade with around 27% in 2005 (1), growing to around 31% in 2010 (1). Africa's share is low, around 3%, (1) but has grown by about 1% (1). North America's share has dropped from just under 15% in 2005 to around 14% in 2010 (1).

Or any other valid point.

(b) Farmers paid fair wage for their work (1) and safer working conditions promoted (1). Money from fair trade can be used to improve services in local communities (1) such as schools and clinics (1) which improves standard of living (1). More money goes directly to farmer, cuts out middlemen who take some of profits for themselves (1). Farmers receive guaranteed minimum price so are not affected as much by price fluctuations (1). Fair trade encourages farmers to treat workers well and to look after environment (1). Often fair trade farmers are also organic farmers who do not use chemicals on crops so protect environment (1). Health care services and education programmes available and tackle problems of HIV/AIDS(1).

Or any other valid point.

15. (a) USA has six out of ten most popular tourist attractions in world including Niagara Falls and Disneyland (1). Most visited tourist attraction is Times Square in USA with 35 million visitors per year (1). Washington D.C. is second most popular tourist

destination with 25 million visitors (1). Trafalgar Square is most popular tourist area in Europe (1). Notre Dame and Disneyland in Paris are most visited attractions in France with 12 million and 10.6 million visitors a year (2). Disneyland Tokyo is most visited attraction in Asia (1). Four out of top ten most popular tourist destinations are Disneyland/Disneyworld parks located on 3 different continents (1).

Or any other valid point.

(b) **People (positive):** Local people employed to build tourist facilities eg hotels (1) and work in restaurants and souvenir shops (1). Employment opportunities allow locals to learn new skills (1) eg obtain foreign language (1) and earn money to improve standard of living (1). Services improved and locals can benefit by using tourist facilities such as restaurants and water parks (1). Better employment opportunities increase the local governments' revenue as wages are taxed (1) so can invest in schools, healthcare and other social services (1). Locals can experience foreign languages and different cultures (1) and can benefit from improvements in infrastructure eg roads and airports (1).

People (negative): Tourist-related jobs are usually seasonal therefore some people may not have income for several months (1) eg at beach and ski resorts (1). Large numbers of tourists can increase noise pollution and upset peace and quiet (1). Local people may not be able to afford tourist facilities as visitor prices are often higher than local rates (1). Tourists can conflict with local people due to different cultures and beliefs (1). There is additional sewage from visitors which increases risk of diseases like typhoid and hepatitis (2).

Environment (positive): Appearance of some areas can be improved by modern tourist facilities (1). Some tourists are environmentally conscious and can have positive impact on landscape by donating money to local projects which help protect local wildlife (1) eg nature reserves (1). Tourist beaches cleaned up to ensure safe for people to use (1) through initiatives like Blue Flag (1). Seas become less polluted as more sewage treatment plants built to improve water quality (1).

Environment (negative): Land lost from traditional uses such as farming and replaced by tourist developments (1). Traditional landscapes/villages spoiled by large tourist complexes (1). Air travel increases carbon dioxide emissions and contributes to global warming (1). Traffic congestion on local roads increases air and noise pollution (1). Tourist facilities such as large high-rise hotels and waterparks spoil look of natural environment (1). Litter causes visual pollution (1). Increased sewage from tourists can cause water pollution (1). Polluted water damages aquatic life and habitats (1).

Or any other valid point.

16. (a) In April 2014 were few cases of Ebola in Africa. By October 2014 were almost 2500 cases in Liberia (1). In Sierra Leone were almost 1,200 cases by October 2014 (1). In Guinea were around 800 cases by October 2014 (1). In Liberia cases rose rapidly from around 250 in August 2014 to around 2500 by October 2014 (1). Sierra Leone witnessed rapid increase in cases from around 500 cases on October 1st 2014 to almost 1200 by mid October 2014 (1).

Or any other valid point.

(b) **Heart Disease:** Lifestyle factors are main cause of heart disease. Many people do not take enough physical exercise which is necessary to keep heart healthy (1). In developed societies many people take car or use lift rather than walking/taking stairs (1). Poor diet leads to heart disease (1). Too much saturated fat can cause hardening or blocking of arteries (1). Many people do not eat enough fruit or vegetables, this can contribute to heart disease (1). Eating too much processed food, with high salt content can also contribute to heart disease (1). Smoking can increase risk of heart disease (1). High stress levels also contribute to heart disease (1). Possible effects of hereditary factors (1).

Cancer: Unhealthy lifestyle is root cause of about third of all cancers (1). Smoking causes almost all lung cancer (1). Poor diet has been linked to bowel cancer, pancreatic cancer and oesophageal cancer (1). Heavy drinking may be a factor in development of cancer (1). Some people may be genetically predisposed to some cancers, eg breast cancer (1). Too much exposure to sun can cause skin cancer (1). Obesity has also been linked with increased cancer risk (1).

Asthma: Infections such as colds or flu affect lungs and narrow airways, making asthma worse (1). Allergic reactions to dust mites in home can cause asthma (1). Pollen from plants outside can cause asthma (1). Traffic fumes in polluted towns and cities can cause asthma (1). Cigarette smoke can cause asthma (1). Asthma can be caused or made worse by damp conditions in home (1). In cases of severe dampness, mould spores may make asthma worse (1).

Or any other valid point.

NATIONAL 5 GEOGRAPHY 2018

1. (a) 827694 – arch (1)

812681 – stack (1)

843662 – cliff (1)

(b) Sand spits (long narrow ridges of sand or shingle) form where the coastline changes direction (1). Longshore drift transports sand (1) and deposits it in a sheltered area (1). Deposited sand builds up over time until it is above sea level (1). This deposition continues until the beach extends into the sea to form a spit (1). Sand spits can also develop a hooked or curved end due to a change in prevailing wind/wave direction (1). Mud flats or salt marsh can develop in an area of calm water behind the spit (1).

Or any other valid point.

2. (a) 893618 – ox-bow lake (1)

883627 – v-shaped valley (1)

895589 – meander (1)

(b) In its middle and lower course, a river rarely flows in a straight line resulting in water flowing from side to side – meandering (1). The water flows faster on the outside and erodes the outside bend of the river channel to form a river cliff (1). This wearing away of the river banks by the river's load is called corrasion (1). Hydraulic action also takes place where water gets into small cracks forcing pieces to break off the river bed and banks (1). The river flows more slowly on the inside bend and deposits some of its load to form a river beach or slip-off slope (1). Over time continuous erosion of the outer bank and deposition on the inner bank forms a meander in the river (1).

Or any other valid point.

3. A = forestry (1)
   B = Halladale River (1)
   C = electricity transmission lines (1)
4. **For Recreation and Tourism:**
   There are a number of suitable places for people to stay. There are buildings such as Armadale House (790638) (1) which could offer bed & breakfast (1). Tourists would also be able to stay at the caravan & campsite at Melvich (887641) (1). There are a number of nice beaches which tourists would enjoy visiting such as at Strathy Bay (8366) and at Armadale Bay (7964) (1). There would be nice sea views for visitors to enjoy from Strathy Point at 828697 (1) and they might also be able to go fishing from the jetty at 831678 (1).

   For Renewable Energy:
   This area could be suitable for wind turbines as there are a number of hilly areas where there would be stronger winds (1). It is also a coastal area where winds tend to be stronger (1). There are a number of tracks which would provide access to hilly areas such as south of Bowside Lodge in 8360 (1). As it is a coastal area, it might be suitable for offshore wind power or wave power (1). An exposed area like this in the north of Scotland is likely to have stronger winds and therefore bigger waves (1). The pier at 883657 would provide local access for boats to maintain marine renewable devices (1). Strathy forest has the potential to provide wood for biomass (1).

   For Forestry:
   The land is not very flat, quite wet in places and so is not good for farming but trees can still be planted here (1) and they clearly grow quite well in this area as there are a number of coniferous plantations already (1) such as at Strathy Forest (8261) (1). Villages such as Melvich and Portskerra might be able to provide a labour force for maintenance and felling (1).

   Or any other valid point.
5. Answers will vary depending on the features chosen:

   e.g. A number of conflicts have emerged around the Dorset coast.

   Strategies include the increase in use of cycle routes and train lines which has helped to reduce traffic congestion on coastal roads (1). Giving the area World Heritage Site status emphasises its worldwide importance helps to protect the coastline (1). Turning a former abandoned quarry into the Townsend Nature Reserve has helped to protect wildlife, flora and fauna (1). It has also been designated a site of special scientific interest (SSSI) which has 7 different species of orchid (1). Marram grass is widely planted to conserve coastal vegetation and reduce the effects of human and physical erosion (1). Lulworth Estate has provided a large car park which has had the effect of reducing some of the parking issues (1). Charges for the car park are spent on improving services in the area which benefit both locals and visitors (1). A bus service has been provided from the nearest train station to encourage visitors to leave the car behind (1). A roundabout has been built at the car park entrance to allow traffic to turn and reduces congestion (1). Lulworth Estate also plans to screen the holiday park to reduce the visual impact on the landscape (1).

   Footpath erosion has been resolved be placing limestone cobbles on paths to make them more durable (1). On steep descents wooden steps have been included to prevent further erosion (1). Reseeding and re-routing paths has protected particularly worn areas (1). Lulworth has no bins to encourage tourists to take litter home (1) and the local estate uses funds from the car park to educate and provide guided tours for tourists (1).

   The MoD have agreed to avoid using the coast on the busiest days of the year: this reduces the impact on tourist experience although there are times when the coast remains restricted (1). The MoD has provided signage to inform visitors when the coastline is closed and which particular areas of paths cannot be used (1). The firing range reduces its practice during peak times to limit the noise from the range (1). The MoD argue that restrictions on access has preserved the natural beauty of the area (1).

   Or any other valid point.
6. **Latitude:** places in southern England are warmer because they are nearer the Equator (1); temperatures generally decrease the further north you go because the sun's rays are less concentrated further away from the equator (1). It's colder in northern Scotland because there is more atmosphere for the sun's rays to pass through (1)

   Altitude/Relief: upland areas are colder as temperatures decrease 1°C for every 100 metres gained in height (1) and wind speeds increase as altitude increases which can affect temperature (1).

   Aspect: in the northern hemisphere south facing slopes can be warmer because they face the sun (1). North facing slopes are shaded from the sun and are therefore cooler (1).

   Continentality/Distance from the sea: places closer to the sea have warmer/milder winters and cooler summers because oceans heat up slowly in summer and cool slowly in winter (1) oceans act as 'thermal reservoirs' (1) whereas places further inland have a greater annual range in temperature due to distance from the effects of the oceans (1). The North Atlantic Drift keeps the temperatures warmer on the west coast than on the east coast of the UK (1).

   Or any other valid point.
7. Due to the approach of the warm front, Stirling's air pressure will fall (1) cloud cover will increase (1) and steady rain will fall (1). Winds will be stronger as the isobars are closer together (1). Because Stirling will be in the warm sector of a depression, temperatures will rise (1) and it will be mild with some cloud cover and occasional showers (1). Due to the cold front arriving, cloud cover will increase (1) with cumulonimbus clouds bringing heavy rainfall to Stirling (1). Temperatures will drop as the cold front passes over (1). As the front begins to move away, the sky will become clear (1) rain will stop (1) air pressure will begin to rise (1).

   Or any other valid point.
8. A. From the public telephone in Henwood (4702) to the school near Rose Hill (5303) 6.25 km. (1)

   B. From Forest Farm (5410) to the church in Stanton St John (5709) 3.75 km. (1)

   C. From Waterperry Gardens (6206) to the College (5502) 8.25 km. (1)

9. (a) Area X – 5106 is the CBD as it has old churches (1) main roads leading to the CBD (1) there are museums here (1). Tourist Information centre (1). Bus station (1)

Area Y – 5502 is a modern suburb as it has a modern street pattern (cul-de-sacs) (1). It is located at the edge of the city as would be expected of the suburbs (1) there are two schools nearby for children of families living in the nearby housing areas (1) there are few main roads, only B class and minor roads (1). There is more open space than would usually be found in either the CBD or inner city (1).

Or any other valid point.

(b) **Advantages**
The land is flat, so easier to build on (1). There is reasonable flat land nearby for expansion or car parking if needed (1). Oxford is nearby, so there is a market for the shopping centre (1). People living in nearby areas, such as New Headington could provide a workforce (1). The A40 is close by to provide easy transport to the area (1). The land is on the outskirts of Oxford so should be cheaper to buy (1). Traffic congestion is also likely to be less of a problem outside of the CBD (1).

Disadvantages
There is a river running through Area Z and this could limit the land available for the development or increase building costs (1). Home Farm (GR 541100) may object to the plans (1). There is a small forest (537099) which would cost money to clear or may cause objections to be raised (1).

Or any other valid point.

10. Answers will vary depending on case study chosen.

Much of Glasgow's housing stock was run down and in need of repair so the government invested money to renovate the old tenements by putting in new windows, bathrooms and double glazing (1). Some of the poorer housing/tower blocks were pulled down (1) and replaced by new housing to improve living conditions in regenerated areas like Glasgow harbour (1). Old grid iron streets plus increased car ownership caused congestion problems so new transport links like the Partick Interchange was built (1). To try to improve unemployment the government invested in the service sector with many jobs being created in call centres (1). Small industrial units replaced the old heavy industries improving employment in the area (1). Tourism has been encouraged with many new hotels appearing in the gap sites left by the demolished factories (1). Improvements made to the environment by landscaping and improving docklands (1).

Or any other valid point.

11. In the Lower Ganges Valley, India, new technology such as tractors allows farmers to increase speed and efficiency (1) which provides better profits for some farmers (1). This money can then be used to improve the overall standard of living of the farmers (1). There is less physical work for people (1) but fewer jobs available (1). This can lead to rural depopulation in some areas such as people leave to find work (1). Machines are expensive and not all farmers can afford them, leading to inequality (1).

The use of irrigation channels can allow two to three harvests a year instead of one, this increases profits for farmers (1). However, as the land is constantly in use the soil quality becomes poorer over time (1).

The increased use of machinery and chemicals has created new industries and jobs, e.g. mechanics to fix tractors (1).

The introduction of GM crops can give the farmer a more reliable harvest as the seeds are designed to resist disease (1). Crops can be grown in adverse conditions, e.g. with less water, ensuring a better food supply for the people (1). However, the increased use of fertilizers and pesticides can damage the environment if they get into the water (1). Farmers become reliant on multi-national companies (1).

Or any other valid point.

12. (a) In 2015 India, China and the USA had the highest Gross National Income (GNI) in the world (1). The GNI of the USA was greater than $4.93 trillion (1). Most African countries for which we have data, had a GNI less than $1.06 trillion (1). The GNI for Brazil was between $2.75.06 – $4.93 trillion (1). Most other South American countries earned less than $1.06 trillion (1). Scandinavian countries all had a GNI of less than $1.06 trillion (1). Germany had a GNI of between $2.75 – $2.93 trillion (1). The UK's GNI was between $1.06 – $2.75 trillion (1).

Or any other valid point.

(b) Answers will vary depending on choice.

If number of people per doctor chosen:
A high number of people per doctor shows a lack of healthcare provision (1). The more people per doctor the less developed a country will be because there isn't enough money to educate them (1). Developing countries often have a poor education system and lack of universities to train qualified doctors (1). Governments in developing countries cannot afford to keep hospital stocked with adequate provisions (1).

If number of births per 1,000 women per year is chosen:
The lower the number of births per women the more developed a country will be because there is a low infant mortality rate and women do not need to have 'extra' children to ensure some survive (1). Children are not needed to work on the land so birth rates are low (1). Contraception is widely available and family planning clinics allow women to plan for a baby (1). Sex education in schools helps to prevent unwanted pregnancies (1).

If percentage of people working in agriculture is chosen:
The lower the percentage of people working in agriculture the more developed a country will be because most people work in factories or services (1). Developed countries have fewer people working in farming because they can afford to import food from other countries (1). People work in mainly secondary and tertiary industries as there is more money to be made in these sectors (1). More people work in agriculture in the developing world because of the lack of mechanisation (1).

Or any other valid point.

13. (a) In 1990 carbon dioxide accounted for around 24,000 million tonnes of emissions. This had risen to around 34,000 in 2010 (1).

In 1990 methane + nitrous oxide accounted for around 10,000 million tonnes of emissions. This had risen to around 12,000 in 2010 (1).

The change in carbon dioxide emissions is more than methane + nitrous oxide or nitrous oxide with a change of around 10,000, compared to 2,000 for methane + nitrous oxide (1). Greenhouse gasses have increased from 1990–2010 (1) from 34,000 million tonnes to 46,000 million tonnes (1).

Or any other valid point.

(b) **Physical Causes:**

Fluctuations in solar activity over time can increase or decrease global temperatures (1).

The Little Ice Age of 1650–1850 may have been caused by a decrease in solar activity (1).

Volcanic eruptions can impact on global temperatures as large quantities of volcanic dust in the atmosphere shield the Earth from incoming insolation which lowers global temperature (1). For example, the eruption of Mount Pinatubo in 1991 caused a dip in global temperatures when 17 million tonnes of sulphur dioxide were released into the atmosphere (1). This reduced global sunlight by 10% and resulted in a 0.5% temperature decrease globally (1).

Large eruptions may also enhance the greenhouse effect and lead to global warming in some instances (1).

Milankovitch cycles or variations in the tilt and/or orbit of the Earth around the Sun affect global temperature (1). More tilt means warmer summers and colder winters, less tilt has the opposite effect (1).

Changes in oceanic circulation such as the periodic warming (El Nino) and cooling (La Nina) of areas of the tropical Pacific Ocean can impact on global temperature (1).

Melting permafrost from Arctic areas can release large quantities of the greenhouse gas, methane (1). This exacerbates the natural greenhouse effect, increasing global temperatures (1).

Human Causes:

The burning of fossil fuels produces carbon dioxide which leads to global warming (1).

Car exhausts, nitrogen fertilisers and power stations all produce nitrous oxide which increases the amount of greenhouse gasses in the atmosphere (1).

Worldwide deforestation also increases carbon dioxide levels, by reducing the storage of carbon (1).

CFCs found in fridges, air conditioning and aerosols contribute to global warming (1).

Increases in rice production and cattle farming contribute to atmosphere pollution (1). Because methane is a stronger greenhouse gas, small increases have a larger impact (1).

Or any other valid point.

14. (a) **Deforestation levels:**

Deforestation rose from just under 1.2 billion hectares in 1900 to around 1.8 billion hectares in 2010 (1). Deforestation levels have steadily increased (1). From 1900 to 1980 deforestation increased from around 1.2 billion to 1.6 billion hectares (1) a difference of 0.4 billion (1).

Population:

World population rose from around 0.9 billion in 1900 to just under 7 billion in 2010 (1). Population increased relatively slowly from 1900 to 1950 from around 0.9 billion to 1.1 billion (1).

Or any other valid point.

(b) **Rainforest:**

Plants such as fan palms have large leaves that are good for catching sunshine and water (1). The leaves are segmented, so excess water can easily drain away (1).

Rainforests have a shallow layer of fertile soil, so trees only need shallow roots to reach the nutrients (1). However, shallow roots can't support huge rainforest trees, so many tropical trees have developed huge buttress roots (1). These stretch from the ground to two metres or more up the trunk and help to anchor the tree to the ground (1).

Lianas are woody vines that start at ground level, and use trees to climb up to the canopy where they spread from tree to tree to get as much light as possible (1).

Strangler figs start at the top of a tree and work down. Gradually the fig sends aerial roots down the trunk of the host, until they reach the ground and take root (2). The figs branches will grow taller to catch the sunlight (1) and invasive roots rob the host of nutrients (1). Eventually the host will die and decompose leaving the hollow but sturdy trunk of the strangler fig (1).

Some plants grow thick leaves with drip tips and waxy surfaces to allow water to drain quickly to prevent rotting (1).

Some plants called 'epiphytes' get food from the air and water, and their roots hang in the air, e.g. orchids (1). Trees grow fast and straight to compete for sunlight (1).

Any other valid point.

Tundra:

Successful plants in the tundra are low growing, compact and rounded in order to help protect from the wind (1). Many grow close together for added protection from the weather (1).

The trees that can survive in the tundra are often small (1) and the snow acts as insulation for the trees and helps them stay warmer during the winter months (1).

During winter months, many plants go dormant to tolerate the cold temperatures (1). By going dormant during the winter, plants are able to save energy and use it during more favourable conditions, like the warmer summer months (1).

Plants grow rapidly during the short summer season, and they flower more quickly (1).

The flowers of some plants increase their heat efficiency by slowly moving during the day to position themselves in a direction where they can catch the most rays from the sun (1). Some plants have cup shaped flowers to trap the sun (1). Other plants have protective coverings, such as thick woolly hairs, that help protect them from wind, cold and desiccation (1).

A small leaf structure is another physical adaptation that helps plants survive. Plants lose water through their leaf surface. By producing small leaves the plant is more able to retain the moisture it has stored (2).

Cotton grass has narrow leaves helping to reduce transpiration (1) its dense flower heads reduce heat loss and darker leaves help absorb energy from the Sun (2).

Or any other valid point.

15. (a) There was a general rise, before a spike in 1995, resulting in around $155 million (1). There was a drop to $30 million in 2001 (1), before rising to $220 million in 2005 (1). There was a high of $360 million in 2011, before it dropped to $150 million in 2012 (1). Overall, there was a rise in damage cost from 1990 to 2012 (1).

Or any other valid point.

(b) Answers will vary depending on the case study chosen.

Scientists can monitor seismic activity. Tremors can give warnings as to an imminent eruption (1). If people are warned they can evacuate (1). Scientists successfully predicted the eruption of Mount St Helens in 1980 by measuring the frequency of earthquakes on the mountain. This enabled many locals to escape to safety (1).

Scientists can also monitor gas emissions, such as sulphur dioxide to predict an eruption (1). However, this is an inexact science and scientists can rarely predict too far in advance. Despite scientists noticing increased tremors and sulphur dioxide emissions before the eruption of Mount St Helens, scientists thought that it might still be a few weeks away (1).

Volcanoes sometimes expel lava bombs before an eruption, this would give the population warning to evacuate (1).

Temperatures around the volcano tend to rise as activity increases. Thermal imaging techniques and satellite cameras can be used to detect heat around a volcano (1).

Volcanoes such as Mount St Helens in the USA and Mount Etna in Italy are closely monitored at all times. This is because they have been active in recent years and people who live nearby would benefit from early-warning signs of an eruption (1). Tilt meters which record changes in the shape of a volcano can also give early warning of an eruption (1).

People living in the shadow of a volcano have emergency plans in place and emergency supplies such as bottled water and tinned food are stockpiled to ensure they have vital supplies to survive in the event of an eruption (2).

In the event of a serious eruption, short term aid in the form of food, medicine and shelter could be sent to the area to treat the injured (1). In the case of Mount St Helens a 5-mile exclusion zone was enforced (1).

When the Pico de Fogo Volcano in Cape Verde erupted in 2014, more than 1,000 people were evacuated from the Cha das Caldeiras region at the foot of the volcano immediately after it first erupted (1). Officials closed the airport as the skies darkened with ash to prevent damage to aeroplane engines (1).

Or any other valid point.

16. (a) The number of people employed in the production Fair Trade flowers and plants has decreased by 1,000 (1) from 51,000 to 50,000 (1). Employees in seed cotton have decreased from 60,000 to 53,000 (1). Cocoa producers have increased by 5,000 (1) from 180,000 to 185,000 (1). The number of employees in the growth of Fair Trade tea has increased by about 65,000 people in one year (1) from 300,000 to 365,000 (1). Tea had the largest increase in the number of employees (1).

Or any other valid point.

(b) Answers will vary depending on case study chosen.

Farmers are paid a fair wage for their hard work producing Fair Trade tea in Kenya (1) and safer working conditions are promoted to prevent accidents and injuries (1). Fair trade also encourages farmers to treat their workers well (1). Farmers receive a guaranteed minimum price for their cocoa in Cote d'Ivoire so they are not affected as much by price fluctuations (1) and can receive some money in advance, so they don't run short (1). More money goes directly to the farmer as the 'middle man' is removed (1). Money from Fair Trade bananas in Latin America can be used to improve services in local communities such as schools and clinics (2) which improves peoples' standard of living (1).

Or any other valid point.

17. (a) There are more endangered world heritage sites located in Africa than in any other continent (1). There is 1 world heritage site in danger in the UK (1). Madagascar has a site that is endangered (1). The Middle East has the second highest number of sites including locations in Afghanistan and Iraq (1). There is also one world heritage site in danger in the USA (1). There are 3 sites in danger in South America (1). Central America has 4 world heritage sites in danger (1). Indonesia has 1 site under threat (1).

Or any other valid point.

(b) National parks are set up to protect fragile environments and to encourage sustainable economic development, including tourism (1). Local guides educate visitors on the importance of conservation (1) and can show them projects where their money is being spent to protect the environment (1). Limited numbers of people are allowed access to eco-tourist areas (1) e.g. in Peru daily numbers are restricted on the Inca Trail (1). Tours must be small-scale so companies have to limit group sizes to lessen environmental impact (1). Eco-tourists must follow local customs and respect local cultures (1) e.g. removing shoes before entering temples in Cambodia (1). Tourists are encouraged to follow the code 'take nothing but photographs, leave nothing but footprints' (1).

Or any other valid point.

18. (a) Answers may include:

HIV/AIDS is most prevalent in developing countries (1). The HIV/AIDS rate is over 10% in South Africa, Namibia and Botswana (1). Kenya and Tanzania have a rate of between 6–10% (1). Mauritania, Mali and Ghana have a rate of between 1–5% (1) South and Central American countries like Brazil and Mexico also have low rates of under 1% (1).

Or any other valid point.

(b) Answers may include:

AIDS is a debilitating disease which means that eventually those infected will not be able to work (1). This lowers productivity and hampers development of a country (1). This in turn leads to fewer jobs and less wealth in a country (1).

The death rate will increase and life expectancy decreases (1).

In areas where AIDS is endemic e.g. South Africa or Uganda, children may be left without parents and brought up by grandparents (1), meaning entire middle-aged populations may be missing from societies (1). Those affected will be mainly in the economically active group so the dependency ratio will increase; there will be less people to support the young and elderly (2).

With more adults ill and unable to work then the economically active population reduces (1), resulting in a shortage of labour (1).

Less food will be produced as less people are able to work the land (1).

There may be a loss of tourist revenue if there are known to be specific problems with disease in the area (1).

The young often become carers, therefore missing out on education (1). There will also be a large number of orphans and dissolved families (1).

Relatives of sufferers may be ostracised by their communities (1).

Lack of staff in schools means that many people don't receive enough education about AIDS (1).

Or any other valid point.

NATIONAL 5 GEOGRAPHY 2019

1. (a) There is a U-shaped valley (5534). (1 mark) There is a misfit stream at 513368 (1). There are several corries on the map such as Coire Odhar at 615404 (1). There is a corrie lochan at 6442 (Lochan nan Cat) (1). There is an arête at 583385 (Cam Chreag) (1). Loch Tay is a ribbon loch (1). There is a truncated spur at 5935. (1).

   Or any other valid point.

   (b) **Stack:**
   Waves attack a line of weakness, such as a fault line, in the headland (1). Types of erosion include hydraulic action and corrasion (1).
   Continuous erosion will open up the crack and it will develop into a sea cave (1). This can happen due to corrasion where stones and pebbles are repeatedly thrown against the cliff face by wave action, wearing it away (1). Further erosion of the cave, often on opposite sides of the headland, will form an arch (1). The base of the arch is attacked by the waves until it eventually collapses (1). This leaves behind a freestanding piece of rock called a stack which is separate from the headland (1).

   Or any other valid point.

2. (a) The river is flowing in a South Easterly direction (1) mainly through a narrow steep sided river valley from 500370 to 545350 (1) but between 510367 to 517366 and 525363 to 537356 the valley is wider and flatter (1). There is a waterfall at 543351 (1). Between 543352 and 570343 the river winds its way through a broad U-shaped valley (1). The river has a large meander at 560344 (1). There are tributaries which join the river at many places, for example 563345 (1). In 5534 the floodplain is more than half a kilometre wide (1).

   Or any other valid point.

   (b) Rainwater absorbs CO_2 from the atmosphere, forming acid rain. In areas of carboniferous limestone, acidic rainwater reacts with the rock and dissolves some of it (carbonation) (1). The dissolved limestone is carried away by running water (solution) (1). The water travels down below ground through joints/bedding planes in the limestone and drips off the roof a cave (1). As it does so, some of the water evaporates leaving a deposit of calcite behind (1). Over time these deposits build up to form an icicle shaped deposit of rock hanging from the roof of the cave (1); these are called stalactites and grow very slowly at a rate of no more than a few millimetres per year (1).

   Or any other valid point.

3. (a) **Advantages**
   There is enough land available so room for ten holiday homes (1). The lack of contour lines shows the land is flat making building easier (1 mark). The site is attractive for visitors as it has good views of the river, mountains and Loch Tay (1). Being so close to Loch Tay would make it easy for guests to go fishing or take part in water sports (1). Being located in the National Park, the area is likely to be popular with tourists, which will be good for business (1).

   Disadvantages
   There is limited access to the site so a better road would have to be built through the countryside (1). Part of this area is within the National Park so planning permission might be harder to get (1). The site is low lying, so in times of heavy rainfall the area might flood (1). Some of the land (for example at 589332) is marshy so would be difficult to build on (1). Some forestry might need to be removed at 584332 which would affect the local wildlife (1). This is a quiet area on the edge of a National Park so additional people in the area will increase the amount of noise pollution and litter, negatively affecting the environment. (2).

   Or any other valid point.

4. Visitors can observe the local wildlife in the National Nature Reserve (1) at 592371 (1). There are many hills and mountains in the area like Ben Lawers, where visitors can go hill walking or mountain climbing (1). There are rivers and lochs where sailing and boating is possible (1). There are many footpaths and tracks in the area, for example at 576350, where visitors can walk through the forest to view the waterfalls (1 mark). There are places of interest like a castle (576337) a stone circle and a fort for visitors interested in history (1). The Falls of Dochart attract visitors for the scenery (1).

5. 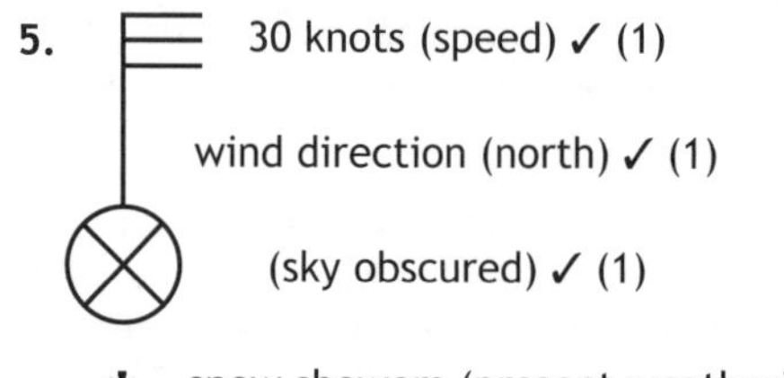

6. As the cold front approaches, the temperature falls as cooler air replaces warmer air on the surface (1). Cold fronts bring periods of heavy precipitation, explaining the heavy showers of rain in the forecast (1). As the cold front moves past Penzance it will bring clearer conditions (1) and there are less clouds allowing periods

of sunshine (1). The isobars are further apart behind the cold front so winds are lighter (1). The direction of the isobars changes after the cold front, explaining why the wind will change direction from SW to W (1).

Or any other valid point.

7. Answers will vary depending upon the case study and land uses chosen.

If Lake District chosen:
The Lake District is a popular tourist area and for some people wind farms spoil the natural beauty of the area (1). Turbines can be highly visible from many directions and can put people off from visiting/returning to the area (1). There are many tourist related jobs in the area, for example gift shops and eating-places like the Kirkstone Pass Inn, and less tourists will adversely affect the economy of the area (1) as well as leading to unemployment (1). New larger turbines are to be built within one mile of the Lake District National Park, which is a protected area (1) and due to their increased height will be much more visible from the mountain tops of the Park, further destroying the views (1). Some areas of the Lake District are designated SSSI sites and the turbines could affect the local nature and wildlife (1). Some areas like Kirkby Moor will no longer be accessible to visitors preventing people going where they want (1). Many people visit the area to take part in quiet pursuits like painting and bird watching but the turbines destroy the character of the area (1). The turbines are noisy and affect the peace and quiet of the local villages (1).

If Cairngorms chosen:
In the Cairngorms, walkers leave gates open and animals escape causing farmers to waste time looking for them (1). Tourists' dogs can worry sheep if let off their lead costing the farmer money in vet fees (1). Stone walls are damaged by people climbing over them instead of using gates/stiles, using up the farmers' time to repair (1).Noisy tourists can disturb sheep especially during breeding season causing them to miscarry (1). Farmers may restrict walkers access at certain times, for example lambing season, preventing tourists going where they want (1). Farm vehicles can slow up tourist traffic on roads adding extra time to journeys (1) and parked cars on narrow country roads can restrict the movement of large farm vehicles (1).

If Yorkshire Dales chosen:
Tourists want to see the beautiful and unusual scenery of the Yorkshire Dales but quarries spoil the natural beauty of the landscape (1). Lorries used to remove the stone endanger wildlife and put visitors off returning to the area (1). This threatens local tourist-related jobs, for example in local restaurants (1). The large lorries needed to remove the quarried stone cause air pollution which spoils the atmosphere for tourists (1). Lorries cause traffic congestion on narrow country roads which slows traffic and delays drivers (1). The peace and quiet for visitors is disturbed by the blasting of rock (1). Some wildlife habitats may also be disturbed by the removal of rock (1). As well as visual and noise pollution, quarries can produce a lot of dust which is unpleasant for nearby residents and visitors (1) and can be spread even further by lorries leaving the quarries (1).

Or any other valid point.

8. A (new industry) = 9004 (1)
B (inner city) = 9101 (1)
C (new housing) = 8699 (1)
D (CBD) = 9198 (1)

9. The contour lines are far apart indicating flat land which is easy to build on (1). The site has room around it for expansion and large car parks (1). Land is cheaper away from the CBD so more room for lower density and large modern buildings (1). Near main roads, for example A4124 for transport of materials and products (1). The Essington Canal could be used, or may have been used in the past, to transport goods (1). Close to large urban area for labour supply and potential customers (1). Further out from the city centre away from congestion and pollution (1) and surrounding open space makes a more pleasant working environment (1). Workers can relax during breaks in the nearby leisure centre (1), for example 937996 (1). The University of Wolverhampton (at 917988) can provide skilled workers. (1)

Or any other valid point.

10. The ring road around the CBD allows traffic to by-pass the busiest part of the city (1). There are numerous roundabouts and sections of dual carriageway to help traffic flow better (1). The bus station provides public transport reducing number of cars in CBD (1). The train station offers an alternative form of public transport reducing the number of cars in the city centre (1). Park and Ride stations allow drivers to park outside the centre (1), for example 933983 (1). There is a tramway and tram stations providing easy access to Wolverhampton from the south-east (1). There are multiple routes into the centre of Wolverhampton so all the traffic is not using the same road especially at rush hour (1). There are traffic-free cycle routes along the canal towpath (for example at 918991) giving a safe way for commuters to bike into the city rather than using their cars (1).

Or any other valid point.

11. **If Rio de Janeiro chosen:**
Wooden shacks have been upgraded to permanent dwellings with some services (1), for example clean-piped water has been provided to help reduce the spread of diseases (1). Self-help schemes are where local people are provided with materials like bricks to upgrade their homes (1). Some prefabricated houses have been built in Rocinha by the Brazilian Government (1) with basic facilities like toilets and electricity (1). The residents have been given legal rights to the land where their house is built (1). Roads have been built/improved in the favela allowing services like rubbish collections to take place (1). In some favelas cable car systems have been constructed to improve transport for residents (1). There have been some schools and health clinics provided for residents (1). Some charities have also donated money to help improve the standard of living of people in slum housing (1), for example by providing computers in schools (1). Security has been improved by having more police patrols (1) which have helped to reduce drugs-related crime (1).

Or any other valid point.

12. **Advantages**

Diversification
Helps to boost the farmers' income when they use other ventures such as farm shops (1). Farmers become more independent and less reliant on subsidies (1). Visiting a farm means people experience rural landscapes and outdoor activities (1). The farmer makes an income from non-farming activities such as accommodation, farm shops, quad-biking, farm attractions, agricultural exhibits, wildlife tours, and country sports (1). Wind farm development on farming land also generates extra income (1).

Government Policy

In the UK, the Department for Food & Rural Affairs (DeFRA) or the Scottish Rural Development Programme (SRDP) supports farming industry by providing subsidies (1). DeFRA regulates policies which improve animal health and welfare regulations (1). Government demands disease control in plants and animals to maintain high standards of produce (1). The Government funds and supports research into agriculture which in turn improves farming practices (1). CAP (Common Agricultural Policy) helps farmers to maintain stable prices and guarantees a steady income (1). Previously, farmers used set-aside land to prevent over-production of certain crops (1). Grants available for environmental improvements (1) such as tree planting or planting hedges in rural land (1).

GM

Genetically modified crops can increase crop yields (1) and improve resistance to disease (1). More tolerant crop varieties could be grown in areas where they couldn't be previously grown (1). GM crops reduce the need for pesticides which is less harmful to insects and bees (1).

New technology

The use of machinery, for example combine harvesters, continues to speed up harvesting and results in the product being delivered to markets fresher (1). Using GPS to manage field operations or animal feeding saves time (1). Computerised water management/irrigation can increase crop production (1). Polytunnels can improve crop yield and quality (1). Chemical fertilisers and insecticides are widely used to improve production on farms (1). Less labour is required which might help to increase farmers' profit margins (1). Drones may be used to survey fields of crops which helps farmers to quickly identify problems (1). Satellite technology/computers used to control the application of fertilisers to particular areas of fields (1) improving yields (1) whilst decreasing the cost and waste, as only the required amounts are delivered to each segment according to the soil quality there (2).

Organic farming

Chemical-free food is grown to meet consumer demands (1). Decreased water pollution due to no chemical run-off (1) protecting aquatic wildlife (1).
Organic produce often sells for a higher price, potentially giving farmers a larger income (1).

Disadvantages

Diversification

Rising tourist numbers in rural areas causes traffic congestion and increased air pollution (1). Dry stone walls are damaged by people climbing over them (1) and footpath erosion damages the natural landscape (1). Increased litter spoils the look of the landscape and can be harmful to wildlife, for example livestock can choke on plastic bags (2). Tourist facilities, for example campsites, detract from the natural look of the countryside (1). Rivers and lochs used for water sports are polluted (1).

Current government policy

It is unclear what impact Brexit will have on UK farmers (1) because they could lose EU funding/grants or face reduced subsidies when they are no longer part of a Common Agricultural Policy (1).

GM

Many people disagree with GM crops, arguing that they may have a negative impact on the natural environment (1). There are health concerns around the use of GM seeds and therefore some people believe they should not be used by farmers (1).

New technology

Increased noise and air pollution from large machinery (1). The cost of buying and maintaining equipment and machinery is expensive (1). Fewer jobs are available and as people become unemployed they move away to find work, leading to rural depopulation (2). Overuse of agricultural chemicals may result in environmental damage (1); ponds and lochs may suffer from eutrophication caused by excess chemical fertilisers (1). Loss of animal habitats, for example hedgerows, which were removed to increase field size and accommodate large machinery (2).

Organic farming

This type of farming is more labour intensive as it takes the farmer more time and effort to grow crops (1). Gaining organic farming status takes time which negatively affects farmers' income at first (1).

Or any other valid point.

13. There is a higher proportion of the population of Bolivia under the age of 15 because birth rates are higher in developing countries where fewer women get the chance of an education (1) and there is less use of contraception (1) and information on birth control is less easily accessed (1). Child mortality rates are higher in Bolivia so people have many children in order to ensure that some survive (1). Many families in Bolivia have lots of children so they can contribute to the family income when old enough and look after them in old age (2). Gender equality in the Netherlands means that there are plenty of career opportunities for women and this reduces birth rates (1). Birth rates are also lower in the Netherlands as women tend to choose to have fewer children later in life (1). In developed countries like the Netherlands people tend to favour lifestyle over larger families (1) and children are expensive to raise, so people have fewer (1). There is a larger proportion of the population over 60 in the Netherlands because more people have the chance of medical treatment than in Bolivia where there is less money to set up health centres and hospitals (2). There is a higher proportion of the population over the age of 60 in the Netherlands as there is clean drinking water so less chance of catching diseases like typhoid (1) and plenty of food to eat, so fewer people die of malnutrition (1). The Netherlands also have pension schemes and facilities to support elderly people, for example care homes, so death rates are lower. (2)

Or any other valid point.

14. (a) As CO_2 concentration increases so does the average global temperature. The amount of CO_2 has increased between 1880–2010 up to 400 PPM (1). The largest increase in the amount of CO_2 was 80 PPM between 1970–2010 (1) increasing from 320 PPM to 400 PPM (1). The average surface temperature has overall increased from 13.8°C to 14.6°C (1). The largest increase in global temperature was between 1895–1900 when it increased by 0.2°C (1). The largest decrease in global temperature was between 1900–1910 when it decreased by 0.25°C (1).

Or any other valid point.

(b) industries and domestic users of energy are encouraged to use it more efficiently through media awareness campaigns (1). People are encouraged to walk, cycle, or use public transport rather than fossil-fuel powered cars (1). Bus lanes and cycle lanes designated to encourage people not to use their car (1). People use smaller more energy-efficient cars or electric cars (1). London Congestion Charge: drivers pay for driving in the Congestion Charge Zone to cut the pollution generated from exhaust fumes (1). Government tax is significantly reduced on vehicles with low CO_2 emissions (1). Encourage people to holiday at home to reduce the number of aircraft journeys taken (especially short-haul flights) (1). Educate people to switch off lights, power sockets, phone chargers and TVs when not in use (1). Recycle and reuse plastics and oil-based products (1). The UK government now levy a charge of 5p for every carrier bag (1). Local councils supply bins to help householders recycle various products (1). Use energy-efficient light bulbs and rechargeable batteries to conserve energy (1). Government grants to help home owners insulate house roofs and use more efficient heating systems (1). Install solar panels on house roof to generate renewable energy (1) or switch to an electricity supplier that supplies green electricity (1). Scientists observe and measure changes in temperature, CO_2 emissions and rising sea levels to monitor the rate of climate change and advise world leaders about the need for action (2). Developed countries switch from fossil fuels to alternative sources of energy in order to reduce the amount of CO_2 in the atmosphere (1). Countries find new types of energy, for example biofuels (1). Industries develop and expand existing sources that are more sustainable than fossil fuels, for example solar, wind and wave power (2). Developing countries reduce deforestation and increase afforestation (2). World summits enable governments to get together and discuss global strategies to try to reduce their use and consumption of carbon-based fossil fuels (1). Many governments signed the Kyoto Protocol/Paris climate agreement, committing them to reducing greenhouse gas emissions (1). The UN climate summit 2014 enabled world leaders to agree actions intended to avert the worst effects of climate change (1). The governments ban the use of harmful substances, for example CFCs (1). The Carbon Credits Scheme is aimed at reducing greenhouse gas emissions by making the polluter pay according to how much pollution they generate (2).

Or any other valid point.

15. (a) As deforestation levels have decreased, the percentage of land being protected has increased in Brazilian Amazon (1). Deforestation levels fell from 30,000 km^2 in 1995 to 18,000 km^2 in 1996 (1). Then decreased further from 18,000 km^2 in 1996 and to 14,000 km^2 in 1997 (1). Deforestation then rose steadily from 1997 to 29,000 km^2 in 2004 (1). After that deforestation levels have decreased to 12,000 km^2 in 2007 (1) with a slight rise to 14,000 km^2 in 2008 (1) and continued to decrease to 4,000 km^2 in 2012 (1) before rising to reach 6,000 km^2 in 2015 (1). Percentage of protected land has risen steadily from 20% in 1995 to 40% in 2010 (1). It then rose sharply to 48% by 2012 (1).

Or any other valid point.

(b) Since 2003, over half of the **Brazilian rainforest** has been designated as national parks/forest reserves/ indigenous lands (1), effectively protecting an area larger than Greenland from intensive logging and agriculture (1). Trees are selected for felling when they reach a particular height, which allows young trees a guaranteed life span and the forest to regain full maturity after around 30–50 years (1). If trees are cut down, they are replaced in order to maintain the canopy and encourage re-afforestation (1). To monitor the pace of deforestation, the Brazilian Space Agency launched its DETER satellite in 2004 (1). This monitoring in Brazil is estimated to have prevented deforestation of 59,000 km^2 of rainforest from 2007 to 2011 (1). The Brazilian government has increased law enforcement on rainforest crime by focusing their efforts on patrolling roads leading into the rainforest (1). Agro forestry – growing trees and crops at the same time lets farmers take advantage of shelter from the canopy of trees and prevents soil erosion (1); the crops benefit from the nutrients from the dead organic matter (1). With the goal of reducing emissions from deforestation, the UN established the Reducing Emissions from Deforestation and Forest Degradation Programme at the 2007 climate summit in Bali (1). Through this programme, industrialised countries with high carbon emissions pay for carbon storage by preserving forests in developing countries (1). This offers developing countries additional economic incentives to preserve their forests and to keep emissions low (1 mark).

Any other valid point.

16. (a) Tropical storms of all categories are more frequent in the Northwest Pacific (1). In both the Northeast and the Northwest Pacific category 1 tropical storms are most frequent (1). The Northwest Pacific has around 85 more category 1 tropical storms than the Northeast (1). In the Northwest Pacific there are at least 100 events in each category of tropical storm, whereas in the Northeast only category 1 tropical storms reach this amount (1). There is a large difference in category 4 and 5 events with the Northwest having around 350 whilst the Northeast has around 90 (1). The largest difference is in category 5 tropical storms, with the Northwest having around 175 compared to the Northeast of around 10 (1), a difference of 165 (1).

Any other valid point.

(b) Tropical storms may form when seeded by sand or dust blown offshore from hot deserts such as the Sahara (1). Tropical storms form over warm oceans which allows heat and moisture to rise upward from the surface of the water and tropical storms need this to fuel their development (1). Sea temperatures have to be approximately 26°C (1). This air moves up and away from the surface so there is less air left near the surface, causing an area of low pressure to form (1). Air from surrounding areas with higher air pressure moves into the low pressure area, then the new air becomes warm and moist and rises, adding more fuel to the system (1). As the warmed, moist air rises it begins to cool off, condense and the water in the air forms clouds (1). The whole system of clouds and wind spins and grows, fuelled by the ocean's heat and water evaporating from the surface (1). The spinning is caused by the rotation of the earth on its axis (1), this is called the Coriolis effect (1). As the storm system rotates faster and faster, an eye forms in the centre (1). Tropical storms weaken when they hit land because they are no longer fed by the energy from the warm ocean waters (1).

17. (a) In 1994 the world's top exporter was the USA but by 2014 it was China (1). In 1994 the USA was the world's largest exporter with approximately $600 billion but by 2004 the USA was the second largest exporter (1) and by 2014 the USA remained the second highest exporter with $1600 billion (1). In 1994 Japan was the third largest exporter with approximately $450 billion but by 2004 it wasn't ranked in the top three (1). Germany ranked as the second highest exporter in 1994 with $500 billion but by 2004 it was the highest exporter (1). By 2014 Germany had dropped down again to third place with $1600 billion (1).

Or any other valid point.

(b) There is a large imbalance in trade between developed and developing countries; this can reinforce differences in wealth between areas such as the EU and Africa (1). Typically, European countries experience a trade surplus while African countries experience a trade deficit (1). African countries export mainly primary products such as tea and cotton for comparatively low prices but import mainly processed/manufactured goods such as vehicles for much higher prices (2), which can result in a trade loss for them (1). This can increase levels of poverty within African countries and cause difficulties for their economy (1). Wealthy European countries profit from selling expensive manufactured goods to African countries (1) and experience a trade surplus, resulting in a greater income and a high standard of living for their citizens (2). Trading blocs such as the EU can have a big impact on world trade patterns (1) because there is free trade between member states (1). Political ideology and sanctions can also impact world trade patterns such as in North Korea or Iran (1).

Or any other valid point.

18. (a) Overall the number of international visitors to Scotland has increased between 2005 and 2017 (1). It has gone up from about 2.4 million people to 3.2 million people by 2017 (1). This is an overall increase of 0.8 million international visitors (1). There are years when it has gone down, such as from 2007 to 2008 when it dropped from 2.7 to 2.5 million. (1) The biggest increase was from 2016 to 2017 when numbers increased from 2.7 to 3.2 million (1). Between 2005 and 2017 international visitor spend has gone up from £1.1 billion to around £2.3 billion (1). The biggest increase in visitor spend was between 2016 and 2017 when it increased from £1.8 billion to £2.3 billion (1). An increase of £500 billion in one year (1). Between 2014 and 2015 international visitor spend dropped from £1.8 billion to £1.7 billion (1).

Or any other valid point.

(b) Eco-tourism aims to promote tourism which conserves the local environment (1). It also encourages an understanding of the culture and way of life in local area (1). Eco-tourism aims to inform and educate tourists about the natural environment of an area (1). It should benefit and provide employment for local people (1) such as travel guides/rangers (1). Money raised from eco-tourism should be used to benefit and conserve the local area (1). It should not impact negatively on the local area in any way (1). Local people should have a say in the promotion of tourist developments (1). Ecotourism can provide funding for the protection of national parks and other natural areas which might not be available from other sources (1). Ecotourism can provide earnings for local communities with few other income – generating options (1). Ecotourism can increase the level of education among travellers, making them more aware of and enthusiastic about conservation (1), for example Kenya's incredible natural diversity is protected in some 50 national parks and reserves (Maasai Wilderness Conservation Trust) (1).

Or any other valid point.

19. (a) Infant mortality is highest in developing countries (1). Europe, North America and Australasia have the lowest rates of child mortality numbers at between 1–25 per 1000 children (1). Most of South America has child mortality rates of between 1–25 apart from Bolivia which has the greatest rates at 51–75 (1), with Guyana and Paraguay having between 26–50 (1). The continent with the worst child mortality rates is Africa with only Libya having between 1–25 (1). Countries like Nigeria, Ghana and Mali have the highest rates between 101–300 (1). Afghanistan has the highest rates in Asia with between 101–300 children (1), followed by Pakistan with between 76–100 (1).

Or any other valid point.

(b) **If heart disease chosen:**
The incidence of heart disease can be reduced through education about positive lifestyle choices. These might include taking regular exercise to help maintain a healthy heart (1) and avoiding too much fatty food to help reduce the build-up of cholesterol in arteries (1). Eating healthily and maintaining a healthy weight will help to reduce strain on the heart (1). As people become more aware of the risks, more people choose to follow healthy lifestyle options in order to reduce the risk of suffering from heart disease (1). Treatment for people with heart disease usually includes the use of different types of medicines such as Rivaroxaban and Warfarin (1) which are designed to thin the blood to reduce the chance of blood clotting (1). Other medicines called statins help to reduce cholesterol, reducing the risk of a heart attack (1), while others are called beta blockers which help to slow down the heart rate, reducing strain on it (1). Sometimes heart surgery such as a bypass operation is required to repair a damaged part of the heart (1), stents can be used to widen narrowed blood vessels (angioplasty) (1) or in the most severe cases patients may be given a heart transplant because their own heart is too weak or damaged (1).

If cancer chosen:
Treatment for cancer may include surgery to remove a tumour (1). Radiation can be used to target a tumour and either destroy it or at least reduce its size (1). Chemotherapy is a common treatment for different types of cancer – this uses drugs to target tumour(s) throughout the body (1) and helps to slow down their growth or destroy them completely (1). Health campaigns help to educate people about the healthy lifestyle choices which they can make to

reduce the risk of contracting cancer. These include stopping smoking as there is a proven link between smoking and lung cancer (1). Choosing to eat a healthy diet with plenty of fresh vegetables and fruit for example, is better for you than lots of processed foods, some of which have been associated with a higher cancer risk (1). There are many screening programmes in operation now where everyone in certain age groups is offered a test to check whether or not they have certain types of cancer, such as breast or bowel cancer (1). If cancer is caught early in this way it is much more likely to be curable (1) and so the death rate from these types of cancer has been reduced by earlier detection (1). People are also encouraged to check themselves for any suspicious lumps and to see their doctor if they find one, which can also help to increase survival rates (1). As obesity is one of the biggest factors increasing the risk of cancer, there are nationwide campaigns aimed at encouraging people to maintain a healthy weight (1).

If asthma chosen:
Treatment for asthma usually involves the use of an inhaler which helps to deliver medication directly into the airways when breathing in (1). Certain types of inhalers (reliever inhalers) aim to relax the muscles around narrowed airways, allowing breathing to become easier and more regular (1). Other types of inhaler (preventer inhalers) deliver medication over a longer time to help prevent the onset of an asthma attack (1). Sometimes oral medication such as steroid tablets can be used to treat patients where inhalers are not having the desired effect (1). Other ways of helping to reduce the incidence of asthma involve targeting the environment in which people live. Some people are allergic to house dust mites, triggering asthma, and so reducing the amount of dust in a house can help to prevent this happening (1). Dust can be reduced by regular dusting and cleaning for example, and the use of barrier materials on bed covers and mattresses (1). Making sure that there are no areas of damp or mould in a house can help as breathing in mould spores may increase the risk of asthma (1). Governments can help to reduce the incidence of asthma by reducing air pollution, especially in congested cities, where microscopic particles of pollution from car exhausts for example, have been shown to lead to many breathing difficulties including asthma (1). Methods of doing this might include introducing a Low Emissions Zone (LEZ) for vehicles in the city centre and more environmentally-friendly methods of transport such as walking, cycling or trams (1).

Or any other valid point.

Acknowledgements

Permission has been sought from all relevant copyright holders and Hodder Gibson is grateful for the use of the following:

Image © Richard Roscoe, Photovolcanica (2017 SQP page 15);
Image is reproduced by kind permission of the Fairtrade Foundation © David Macharia (2017 SQP page 16);
Image © jan kranendonk/Shutterstock.com (2017 SQP page 17);
Image © mountainpix/Shutterstock.com (2018 Section 1 page 7);
Image © Martin Kemp/Shutterstock.com (2018 Section 1 page 7);
Image © Christopher Elwell/Shutterstock.com (2018 Section 1 page 7);
Image © chris2766/stock.adobe.com (2018 Section 1 page 7);
Image © cybrain/Shutterstock.com (Elements of this image furnished by NASA) (2018 Section 3 page 17);
Image © Silken Photography/Shutterstock.com (2018 Section 3 page 19);
Image © Hugh Lansdown/Shutterstock.com (2018 Section 3 page 19);
Image © Daimond Shutter/Shutterstock.com (2018 Section 3 page 19);
Image © Bildagentur Zoonar GmbH/Shutterstock.com (2018 Section 3 page 19);
Image © Jack Cronkhite/Shutterstock.com (2018 Section 3 page 19);
Image © nalongsak hoisangwan/Shutterstock.com (2018 Section 3 page 19);
Image © Leonid Ikan/Shutterstock.com (2018 Section 3 page 19);
Image © BMJ/Shutterstock.com (2018 Section 3 page 19);
Image © BerndBrueggemann/stock.Adobe.com (2019 Section 1 page 2);
Image © Sue Hwang/Shutterstock (2019 Section 1 page 4);
Image © chaiviewfinder/Shutterstock.com (2019 Section 1 page 7);
Image © Donatas Dabravolskas/Shutterstock.com (2019 Section 2 page 10);
Image © PPeteG/stock.Adobe.com (2019 Section 2 page 11);
Poster and quote from https://un.org.au/2014/09/25/climate-summit-2014/ © 2014 United Nations. Used with permission of the United Nations (2019 Section 3 page 15);
Image © guentermanaus/Shutterstock.com (2019 Section 3 page 17);
Image © Harvepino/Shutterstock.com (2019 Section 3 page 19);
Ordnance Survey maps © Crown Copyright 2019. Ordnance Survey 100047450.